Kevin Leckey

Probabilistic Analysis of Radix Algorithms on Markov Sources

Kevin Leckey

Probabilistic Analysis of Radix Algorithms on Markov Sources

Südwestdeutscher Verlag für Hochschulschriften

Impressum / Imprint
Bibliografische Information der Deutschen Nationalbibliothek: Die Deutsche Nationalbibliothek verzeichnet diese Publikation in der Deutschen Nationalbibliografie; detaillierte bibliografische Daten sind im Internet über http://dnb.d-nb.de abrufbar.
Alle in diesem Buch genannten Marken und Produktnamen unterliegen warenzeichen-, marken- oder patentrechtlichem Schutz bzw. sind Warenzeichen oder eingetragene Warenzeichen der jeweiligen Inhaber. Die Wiedergabe von Marken, Produktnamen, Gebrauchsnamen, Handelsnamen, Warenbezeichnungen u.s.w. in diesem Werk berechtigt auch ohne besondere Kennzeichnung nicht zu der Annahme, dass solche Namen im Sinne der Warenzeichen- und Markenschutzgesetzgebung als frei zu betrachten wären und daher von jedermann benutzt werden dürften.

Bibliographic information published by the Deutsche Nationalbibliothek: The Deutsche Nationalbibliothek lists this publication in the Deutsche Nationalbibliografie; detailed bibliographic data are available in the Internet at http://dnb.d-nb.de.
Any brand names and product names mentioned in this book are subject to trademark, brand or patent protection and are trademarks or registered trademarks of their respective holders. The use of brand names, product names, common names, trade names, product descriptions etc. even without a particular marking in this work is in no way to be construed to mean that such names may be regarded as unrestricted in respect of trademark and brand protection legislation and could thus be used by anyone.

Coverbild / Cover image: www.ingimage.com

Verlag / Publisher:
Südwestdeutscher Verlag für Hochschulschriften
ist ein Imprint der / is a trademark of
OmniScriptum GmbH & Co. KG
Heinrich-Böcking-Str. 6-8, 66121 Saarbrücken, Deutschland / Germany
Email: info@svh-verlag.de

Herstellung: siehe letzte Seite /
Printed at: see last page
ISBN: 978-3-8381-5150-2

Zugl. / Approved by: Frankfurt, J.W. Goethe Universität, Diss., 2015

Copyright © 2015 OmniScriptum GmbH & Co. KG
Alle Rechte vorbehalten. / All rights reserved. Saarbrücken 2015

Contents

Introduction		i
1 Models and Results		**1**
1.1	Radix Sort	1
1.2	Input Models	3
1.3	Distributional Recursions	5
1.4	Results on Radix Sort and Related Problems	7
1.5	Results on Radix Select	8
2 Applications		**13**
2.1	Radix Sort	14
2.2	Digital Trees	15
	2.2.1 Analysis of Tries	18
	2.2.2 Analysis of PATRICIA Tries	19
	2.2.3 Analysis of Digital Search Trees	21
3 Techniques		**23**
3.1	Asymptotic Analysis	25
3.2	Poissonization and Depoissonization	32
3.3	The Contraction Method	35
	3.3.1 Contraction Method in the Bernoulli Source Model	35
	3.3.2 Generalization to the Markov Source Model	41
	3.3.3 Lipschitz-Continuity in the Asymptotic Analysis	44
4 Moments and Limit Theorems		**47**
4.1	Analysis of the Mean	48
4.2	Analysis of the Variance	50
4.3	Limit Theorems (Contraction Method)	64
4.4	Transfer to Arbitrary Initial Distributions	71
5 The Radix Selection Algorithm		**75**
5.1	Introduction	76
	5.1.1 The Wasserstein Metric	77
5.2	The Quantile Model	78
	5.2.1 Worst Case Behavior	79
	5.2.2 Selection of Quantiles in the Markov Source Model	84
	5.2.3 A Remark on Convergence in $\mathcal{D}[0,1]$	96

5.3	Grand Averages	98
	5.3.1 Transfers from the Quantile Model	99
	5.3.2 Analysis with the Contraction Method	101
	5.3.3 A Remark on the Concentration for Grand Averages	107

6 Conclusions   **109**
    6.1 Open Problems 110

Appendix   **111**
    A.1 Tail Bounds for the Binomial Distribution 111
    A.2 Moment Bounds 112

Introduction

The analysis of algorithms is a field in theoretical computer science and applied mathematics dedicated to the study of space and time requirements of algorithms. A detailed analysis reveals the advantages and weaknesses of different algorithms, making it possible to compare their performance on the same task. This knowledge in return offers computer scientists an efficient way to choose an algorithm for a particular application, avoiding results which in the end would be anything but efficient.

Typically, the theoretical study of algorithms is focused on a parameter which is independent of the particular implementation, such as the number of comparisons made by comparison-based sorting algorithms. These parameters usually depend on the input of the algorithm. In some applications, there is a "worst case" input in the sense that the parameter is maximized for a typical kind of input. For randomized algorithms, there might also be a worst case behavior of the algorithm that maximizes the running time. However, focusing the study solely on the worst case behavior of algorithms for comparison reasons is not always the best choice.

Quicksort is a very famous example for an algorithm that requires $\Theta(n^2)$ comparisons for sorting n elements in the worst case but still, most of the time, beats many other sorting algorithms with a $\Theta(n \log n)$ worst case. Thus, it is reasonable to discuss the running time of algorithms applied to a "typical" input and, for randomized algorithms, to focus on the "typical" running time. Since the number of comparisons made by *Quicksort* is only dependent upon the ranks of the elements that need to be sorted, there are only a finite number of essentially different inputs of n elements, namely the $n!$ different permutations of $1, \ldots, n$. Therefore, a "typical" input may be modeled to be a uniformly chosen permutation of $1, \ldots, n$.

In other applications, such as the non-comparative sorting algorithm *Radix Sort* or the storage of strings in *Tries*, the input has a more complicated influence on the running time. In particular, the running time may change drastically for inputs drawn according to different distributions on $[0, 1]$. For instance, Luc Devroye's paper [8] implies that the expected number of *Bucket operations* performed by *Radix sort* (details given in the next chapter) on n i.i.d. numbers drawn from $[0, 1]$ according to some density f might be either infinite or $n \log_2 n + O(n)$, depending on whether f is very "peaky" or not (the case $n \log_2(n)$ includes all densities with $\int (f(x))^2 dx < \infty$; the first case includes decreasing densities with $\sum_{k=1}^{\infty} (\int_0^{2^{-k}} f(x) dx)^2 = \infty$).

Thus, the choice of the input model is of significant importance in order to obtain reliable predictions about how an algorithm performs in practice. However, a theoretical study of an algorithm usually requires some restrictions on the input model in order to make detailed (and provable) statements about its performance. The work of Luc Devroye on the *Density Model* covers a quite general input model for numbers in the unit interval $[0, 1]$. Since atoms in the distribution of the input should be excluded (in order to avoid the possibility of two equal numbers in the input), the assumption of a density for the distribution of each number is very natural.

However, this thesis is focused on data structures and algorithms that operate on strings which, in many applications, are considered to model words. More precisely, a string is a sequence of symbols drawn from an alphabet Σ. If Σ is ordered, there is an induced lexicographical order on the set of all strings. A very basic model for random strings is given by the *Memoryless Source Model* (for $\Sigma = \{0, 1\}$ also called the *Bernoulli Source Model*), in which each string is considered to be a sequence of i.i.d. symbols drawn from Σ. Obviously, this is not a realistic model for words in practice (e.g. English words) due to the lack of dependence between consecutive symbols. But still, it captures several informative features (and methodical limitations) which may be extended to more realistic models. A step towards a more realistic model is made by the *Markov Source Model*. Within this model, the symbols of each string are considered to be distributed according to a Markov chain on Σ.

The classical *Radix Sort* induces a sorting algorithm on strings, sometimes called *Triesort*. The performance of *Radix Sort* is closely related to tree-like data structures known as *Digital Trees*. The first part of this thesis is dedicated to the analysis of *Radix Sort* and the path length of *Digital Trees* under the *Markov Source Model*. This includes an asymptotic analysis of the mean and variance and a limit law as the number of strings tends to infinity. The second part is focused on *Radix Select*, an algorithm that selects an element of a given rank instead of sorting the entire list. There are three different kinds of models discussed in this thesis, all of them considering strings generated by a Markov source:

- the worst case study considers maximal cost of selecting an element of rank $1, \ldots, n$, (thus, the cost is maximized over the possible ranks; the strings remain randomly generated)
- the *Grand Averages Model* investigates the cost of selecting a uniformly distributed rank (the distribution of the rank is independent of the strings),
- the *Quantile Model* is based on the study of all quantiles $\lfloor tn \rfloor + 1$, $t \in [0, 1]$, as the number n of strings tends to infinity.

Despite the very successful application of complex analytical methods to the moment analysis of random recursive structures, all asymptotic results on mean and variance given in this thesis are based on transfer theorems derived from well known tail inequalities of the binomial distribution. These transfer results are fitted to enable a limit law with the *Contraction Method*, a method introduced by Uwe Rösler in 1991 and successfully generalized and applied to a variety of random recursive structures and algorithms.

A very brief summary of what is known about Digital Trees

More than fifty years ago, in 1963, Don Knuth wrote his "Notes on Open Addressing", starting his pioneer work in a field nowadays called the analysis of algorithms. Since then, a variety of different methods were applied to analyze algorithms, including analytical methods such as *Singularity Analysis, Saddle Point Strategies, Mellin Transforms, Rice Method* and probabilistic methods such as *Renewal Theory, Moment Methods* and the *Contraction Method*.

Many analytical techniques in the analysis of algorithms are based on the work of Ph. Flajolet. His pioneer work in the field of *Analytic Combinatorics* enabled a systematic average case analysis of recursive structures and algorithms with a precision that often cannot be achieved by most of the other methods. A survey on the field of *Analytic Combinatorics* is given in the book [18] of Flajolet and Sedgewick. An introduction into several analytical and probabilistic methods is given in Szpankowski's book [67]. Moreover, there is a very informative article [19] by Fuchs, Hwang and Zacharovas showing how analytical methods may be applied to analyze the variance of a large family of recursive structures and algorithms.

Concerning probabilistic methods, Janson illustrates in his article [38] how results from *Renewal Theory* may be applied to the analysis of *Tries*. Applications of the *First and Second Moment Method* and the *Inclusion-Exclusion Principle* are given in Szpankowski's book [67]. The *Contraction Method*, a method which many results in this thesis rely on, was introduced by Rösler in [59] in order to derive a limit law for the complexity of the Quicksort algorithm. Since then, several applications and extensions of this method were made, cf. [12, 14, 39, 54, 55, 56, 57, 60, 61]. In the context of this method Neininger and Rüschendorf developed in [54] the use of a different kind of metric, replacing the Wasserstein ℓ_p metrics with the Zolotarev metrics. This change allowed them to cover a large class of recursive structures and algorithms (including those with a normal limit) which till then could not be handled by the Wasserstein metrics.

Many of these methods were successfully applied to the analysis of *Digital Trees*. This brief summary contains some of the results on *Tries* and *Digital Search Trees* for different input models:

Memoryless/Bernoulli Sources: Jacquet and Régnier gave several limit laws for *Trie* parameters based on analytical methods, including a normal limit law for the size and external path length in the *Bernoulli Source Model* in [28] and a limit law for the depth and height in [27]. Some of these results were re-derived with the *Contraction Method* in [54]. Moreover, Jacquet and Régnier studied the size of a *Trie* in [30], focused on an asymptotic expansion of the variance.

There are several publications by Szpankowski that deal with *Digital Trees*, not only restricted to *Memoryless Sources*. In particular, he considered the depth and path length of *Tries* for *Memoryless Sources* in [64]. Results on the profile of *Tries* are given by several authors in [26].

Note that the external path length of *Tries* follows the same distribution as the number of *Bucket Operations* performed by *Radix Sort*. Thus, all results on the path length also hold for *Radix Sort*.

In the case of *Digital Search Trees*, Kirschenhofer and Prodinger gave an asymptotic expansion of mean and variance of the depth for a *Symmetric Bernoulli Source* in [40]. Louchard added a limit theorem under the same assumption in [49]. This result was extended to *Asymmetric Bernoulli Sources* in [50]. In fact, it turns out that the limit is gaussian if and only if the source is asymmetric.

The profile was studied for *Symmetric Bernoulli Sources* by Knessel and Szpankowski in [45] and extended to *Asymmetric Bernoulli Sources* in [13]. Moreover, the variance of the path length for a *Symmetric Bernoulli Source* was studied by Fuchs, Lee and Prodinger in [20] (based on a work of Hwang, Fuchs and Zacharovas). A result on *Asymmetric Bernoulli Sources* is given by Hubalek in [24]. Finally, Szpankowski and Louchard derived a gaussian limit law for the path length under the *Bernoulli Source Model* in [36].

Markov Sources: The analysis of the depth of *Tries* under the *Markov Source Model* was done by Jacquet and Szpankowski in [31]. In particular, they showed a gaussian limit law for *Markov Sources* (excluding the *Symmetric Bernoulli Source*) and derived an asymptotic expansion of mean and variance. These results were based on the *Inclusion-Exclusion Rule*.

Moreover, the depth of *Digital Search Trees* was studied for *Markov Sources* in [37]. There seems to be no publication on the path length of *Digital Trees* for *Markov Sources* so far, a gap that was recently filled by [48]. The results in [48] are based on a particular case pertaining to the methods presented in this thesis.

Additional results regarding *Markov Sources* may be deduced from the corresponding results on the more general *Dynamical Sources Model*. These results mainly cover asymptotic expansions of the mean of several tree parameters.

Density Model: The *Density Model* was introduced and analyzed by Devroye. Several results on this model are given in [6, 7, 8].

More General Sources: The *Dynamical Sources Model* is more of a general input model introduced by Vallée in [68]. The analysis of *Tries* under the *Dynamical Sources Model* started with the work of Clément, Flajolet and Vallée in [2], focusing on the asymptotic analysis of the expectation of several parameters, such as height, depth and size. Moreover, there is a recently published work [1] on *Tame Sources* by Cesaratto and Vallée and a result on the depth of *Tries* and *Digital Search Trees* in [25] by Vallée and Hun.

A Remark on Comparison-based Algorithms: Typically, the analysis of comparison-based algorithms is focused on the number of key comparisons. However, since a key comparisons usually involves several bit comparisons, it takes a more detailed study of these algorithms in order to compare them with *Radix Sort*. Some results on the number of bit comparisons performed by some sorting algorithms (including *Quicksort* and *Quickselect*) are given in [3] and [4].

Notation

Throughout this thesis, $\mathfrak{P}$ denotes the set of all probability measures on $\mathbb{R}$. Moreover, $\mathfrak{P}_s \subset \mathfrak{P}$ denotes the set of all probability measures on $\mathbb{R}$ with finite s-th moment for $s > 0$. The following abbreviations are used for some well known probability distributions:

- δ_x denotes the Dirac measure in $x \in \mathbb{R}$,
- $B(n,p)$ is the binomial distribution with $n \in \mathbb{N}$ trials and success probability $p \in [0,1]$,
- $B(p) := B(1,p)$ denotes the Bernoulli distribution with success probability $p \in [0,1]$,
- $\Pi(\lambda)$ denotes the Poisson distribution with mean $\lambda > 0$,
- $\mathcal{N}(\mu, \sigma^2)$ denotes the normal distribution with mean $\mu \in \mathbb{R}$ and variance $\sigma^2 > 0$,
- $unif(M)$ denotes the uniform distribution on M, where M is either a finite set or an interval of finite length.

The distribution of a random variable X is denoted by $\mathcal{L}(X)$. Equality in distribution is abbreviated by $\stackrel{d}{=}$, i.e. for the random variables X and Y

$$X \stackrel{d}{=} Y \iff \mathcal{L}(X) = \mathcal{L}(Y).$$

Moreover, the L_s norm of a random variable X with $\mathcal{L}(X) \in \mathfrak{P}_s$ is denoted by

$$\|X\|_s := \mathbb{E}[|X|^s]^{\min\{1, 1/s\}}, \quad s > 0.$$

Types of convergence: Weak convergence of measures $\mu_n, \mu \in \mathfrak{P}$, $n \geq 1$, is denoted by $\mu_n \xrightarrow{w} \mu$. If $(X_n)_{n \geq 0}$ is a sequence of random variables and X is a possible limit, the following standard notation for different kinds of convergence is used as $n \to \infty$:

- **Convergence in distribution:** $X_n \xrightarrow{d} X$ if and only if $\mathcal{L}(X_n) \xrightarrow{w} \mathcal{L}(X)$.
- **Convergence in probability:** $X_n \xrightarrow{\mathbb{P}} X$ if and only if $\mathbb{P}(|X_n - X| \geq \varepsilon) \to 0$ for all $\varepsilon > 0$.
- **Convergence in L_p:** $X_n \xrightarrow{L_p} X$ if and only if $\|X_n - X\|_p \to 0$.

Big O notation: The following notation is used for real valued sequences $(a_n)_{n \geq 0}$ and $(b_n)_{n \geq 0}$:

$a_n = O(b_n)$ $\iff$ there exist $n_0 \in \mathbb{N}$ and $C > 0$ such that $|a_n| \leq C|b_n|$ for all $n \geq n_0$,

$a_n = o(b_n)$ $\iff$ $|b_n| > 0$ for all but finitely many n and $\dfrac{a_n}{b_n} \to 0$ as $n \to \infty$,

$a_n \sim b_n$ $\iff$ $|b_n| > 0$ for all but finitely many n and $\dfrac{a_n}{b_n} \to 1$ as $n \to \infty$,

$a_n = \Theta(b_n)$ $\iff$ $a_n = O(b_n)$ and $b_n = O(a_n)$.

If $|b_n| > 0$ for all $n \geq 0$, $a_n = O(b_n)$ is equivalent to the existence of a constant $C > 0$ such that $|a_n| \leq C|b_n|$ for all $n \geq 0$.

The typical Bachmann-Landau notation is also used for functions $f : D \to \mathbb{R}$ and $g : D \to \mathbb{R}$ with $D \subset \mathbb{R}$. The limiting behavior of $f(x)$ and $g(x)$ is always meant as $x \to \infty$ if not stated otherwise.

Some more notations. Finally, a small list of frequently used notations:

- $x \vee y := \max\{x, y\}$ denotes the maximum of two numbers $x, y \in \mathbb{R}$.
- $x \wedge y := \min\{x, y\}$ denotes the minimum of two numbers $x, y \in \mathbb{R}$.
- $\Delta a(n) := a(n+1) - a(n)$ denotes the (forward) difference of a sequence $(a(n))_{n \geq 0}$. Note that $\Delta a(m_n) = a(m_n + 1) - a(m_n)$ for another sequence $(m_n)_{n \geq 0}$.
- $\mathcal{D}[0, 1]$ denotes the space of all càdlàg functions $f : [0, 1] \to \mathbb{R}$.
- d_{sk} denotes the Skorokhod distance on $\mathcal{D}[0, 1]$.

Chapter 1

Models and Results

This chapter is dedicated to the presentation of the type of random recursive structures and algorithms which are examined throughout the thesis. *Radix Sort* is a prime example for such a recursive algorithm. The performance of those structures and algorithms depends on the input which may be modeled in different ways. Therefore, there are several random input models summarized in section 1.2 including the most common *Bernoulli Source Model* and the *Markov Source Model*.

Motivated by the analysis of the *Radix Sort* performance under random input, section 1.3 deals with the presentation of distributional recursions covered by the results of this thesis.

These results are presented in section 1.4 including the first order asymptotic of mean and variance and a limit theorem as the size of the input tends to infinity.

Aside from *Radix Sort*, applications such as the study of the path length in *Tries*, *PATRICIA Tries* and *Digital Search Trees* are presented in chapter 2.

The proof of the results rely on some transfer techniques derived in chapter 3 and on the *Contraction Method* introduced in section 3.3. Complete proofs of the results are given in chapter 4.

There are some other results on a selection algorithm called *Radix Select* that are shown in chapter 5. The recursive behavior of *Radix Select* differs from the other structures discussed in this chapter and requires a readjustment of the techniques presented in chapter 3.

1.1 Radix Sort

The classical way to think of *Radix Sort* is as a non-comparative integer sorting algorithm. However, there are several other variants of *Radix Sort* with an input that may either be a list $[u_1, \ldots, u_n] \in [0,1]^n$ of n numbers in the unit interval $[0,1]$ or a list $[w_1, \ldots, w_n]$ of n words with symbols in some finite (ordered) alphabet. For words, the output is a list of words that are sorted in lexicographical order.

Radix Sort on $[0,1]$: The *Radix Sort* algorithm applied to a list $[u_1, \ldots, u_n]$ of numbers in $[0,1]$ works as follows: Fix a base $b \geq 2$ and derive the b-ary expansion of $u_1, \ldots, u_n$, i.e. determine sequences $(u_k^{(i)})_{k \geq 1}$ in $\{0, \ldots, b-1\}$ such that for all $i = 1, \ldots, n$

$$u_i = \sum_{k=1}^{\infty} u_k^{(i)} b^{-k}.$$

Afterwards, distribute $u_1, \ldots, u_n$ into b sublists (also called *Buckets*) according to their first symbol in the b-ary expansion, i.e. divide $u_1, \ldots, u_n$ into lists $L_0, \ldots, L_{b-1}$ such that u_j is in list L_i if and only if $u_1^{(j)} = i$. The elements in the lists are partially sorted in the sense that for any $i < j$, all elements of L_i are smaller than any element of L_j.

The algorithm is iterated in each list L_j with at least two elements by dividing the elements into lists/Buckets $L_{j,0}, \ldots, L_{j,b-1}$ according to their second symbol in the b-ary expansion. The iteration continues until each sublist holds at most one element.

A natural measure for the complexity of the algorithm is the number of *Bucket Operations* where a *Bucket Operation* denotes the placement of an element into a sublist. An example for *Radix Sort* ($b = 2$) and the number of *Bucket Operations* is given in figure 1.1.

For simplicity and due to the fact that a binary representation of numbers is natural in computer science, we only consider *Radix Sort* with $b = 2$ *Buckets*.

Figure 1.1 The *Radix Sort* algorithm on 6 strings represented by the splitting of the lists. Each string may also be interpreted as a binary expansion of a number in the unit interval. The total number of *Bucket Operations* is $6 + 3 \cdot 3 + 3 \cdot 2 = 21$.

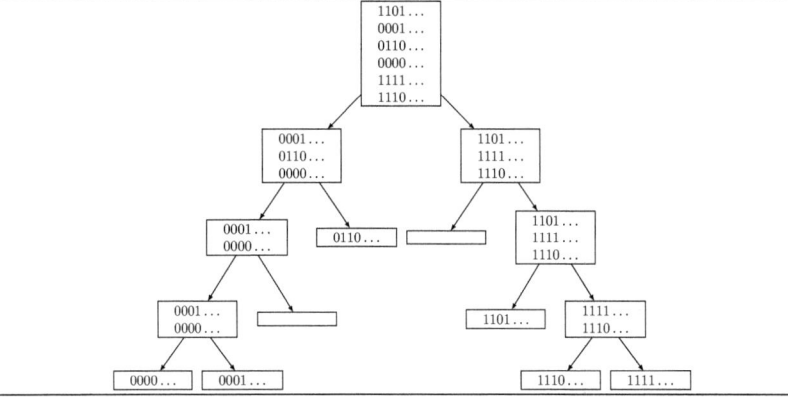

Radix Sort on words: Suppose that Σ is some finite, ordered alphabet, e.g. $\Sigma = \{0, \ldots, b-1\}$ for some $b \geq 2$. Consider a list $[w_1, \ldots, w_n]$ of words/strings of infinite length, i.e.

$$w_j = \left(w_k^{(j)}\right)_{k \geq 1}$$

with $w_k^{(j)} \in \Sigma$ for every $j = 1, \ldots, n$ and $k \in \mathbb{N}$.

The lexicographical order of words is given by: $(v_k)_{k \geq 1} < (w_k)_{k \geq 0}$ if and only if there exists a $k_0 \in \mathbb{N}$ such that

$$v_k = w_k \text{ for all } k < k_0 \quad \text{and} \quad v_{k_0} < w_{k_0}.$$

1.2. INPUT MODELS

The *Radix Sort* algorithm presented for numbers in $[0, 1]$ may also be applied to words by directly taking the words $w_1, \ldots, w_n$ instead of the b-ary expansions of $u_1, \ldots, u_n$.

Setting in this thesis: Throughout this thesis, the input of *Radix Sort* is a finite list $\mathcal{X} = [\Xi_1, \ldots, \Xi_n]$ of n (distinct) strings $\Xi_1, \ldots, \Xi_n$ such that each string is a sequence of symbols taken from the binary alphabet $\Sigma = \{0, 1\}$. The output is a reordered list $[\Xi_{\sigma(1)}, \ldots, \Xi_{\sigma(n)}]$ such that σ is some permutation on $\{1, \ldots, n\}$ and $\Xi_{\sigma(1)} < \Xi_{\sigma(2)} < \ldots < \Xi_{\sigma(n)}$ are in lexicographical order.

The complexity of the algorithm is measured in the number of *Bucket Operations*. A *Bucket Operation* denotes the placement of a string into a sublist (also called *Bucket*).

The strings $\Xi_1, \ldots, \Xi_n$ may either be interpreted as words or as the binary expansion of numbers $u_1, \ldots, u_n$ in the unit interval. Note that the lexicographical order of the binary expansion coincides with the regular order of $u_1, \ldots, u_n$.

The study of the performance requires a model for the input list $\mathcal{X}$. The next section presents some reasonable models to randomly generate a string Ξ. In these models, the list $\mathcal{X}$ is considered to contain n independent copies of Ξ.

1.2 Input Models

All input models presented in this section describe how a single string $\Xi = (\xi_j)_{j \in \mathbb{N}}$ is generated. Strings $\Xi_1, \ldots, \Xi_n$ in these models are considered to be independent and identically distributed with the same distribution as Ξ.

However, there are some other models where $\Xi_1, \ldots, \Xi_n$ are not independent. An example for a model with dependent strings is the *Suffix Tree Model*. Here, one considers strings $\Xi_1, \ldots, \Xi_n$ where Ξ_i is the i-th suffix of a given string Ξ. Some details on suffix trees can be found in [11, 33, 66].

The *Bernoulli Source Model* is one of the simplest stochastic models for strings. In this model, the strings are independent and identically distributed as a string Ξ that is generated as follows:

Definition 1.2.1. *A (random) string* $\Xi = (\xi_j)_{j \in \mathbb{N}}$ *is generated by a* Bernoulli Source *with success probability* $p \in (0,1)$ *if* Ξ *is a sequence of independent and identically distributed random variables such that* $\mathbb{P}(\xi_j = 1) = p = 1 - \mathbb{P}(\xi_j = 0)$ *for all* $j \in \mathbb{N}$. *A Bernoulli Source is called* symmetric *if* $p = \frac{1}{2}$.

Note that, depending on p, one expects independent copies $\Xi_1, \ldots, \Xi_n$ of Ξ to either have long common prefixes (if p is close to 0 or 1) or to be well balanced (if p is close to $\frac{1}{2}$) in the sense that each split in the *Radix Sort* algorithm leads to two sublists of nearly equal size. One of the key parameters that captures this effect of p is the *Source Entropy* given by

$$H_{Ber(p)} = -p \log p - (1-p) \log(1-p). \tag{1.1}$$

Most of the work dealing with *Radix Sort* and related problems is done under this model, see e.g. [28, 16, 41, 42, 46, 51, 19] and the references therein for some results on *Digital Trees*.

However, a *Bernoulli Source* is not very realistic in modeling words of a given language due to the lack of dependence between the symbols in each string. One step towards a more realistic model is to generate each symbol ξ_j depending on the value of the previous symbol. The *Markov Source Model* allows such a dependence. There, each of the n independent strings is generated as follows:

Definition 1.2.2. *Let $P = (p_{ij})_{i,j \in \{0,1\}}$ be a stochastic matrix and $\mu = \alpha \delta_0 + (1-\alpha)\delta_1$, $\alpha \in [0,1]$, be a probability distribution on $\{0,1\}$. A (random) string $\Xi = (\xi_j)_{j \in \mathbb{N}}$ is generated by a Markov Source with initial distribution μ and transition matrix P if $(\xi_j)_{j \in \mathbb{N}}$ is a Markov chain with initial distribution μ and transition matrix P, i.e. if for any $n \in \mathbb{N}$ and $(x_1, \ldots, x_n) \in \{0,1\}^n$*

$$\mathbb{P}\left(\bigcap_{j=1}^{n}\{\xi_i = x_i\}\right) = \mu(\{x_0\}) \prod_{j=2}^{n} p_{x_{j-1}x_j}.$$

Note that not all choices for P are reasonable. In fact, the presented version of *Radix Sort* only works on a list of distinct strings (this is even more important for the other structures presented in chapter 2). Hence, in any reasonable stochastic model, two independent copies Ξ_1 and Ξ_2 of Ξ should satisfy

$$\mathbb{P}(\Xi_1 = \Xi_2) = 0.$$

This requires that the distribution $\mathcal{L}(\Xi)$ of the entire string has no atoms. Therefore, stochastic matrices P with $p_{00} = 1$ or $p_{11} = 1$ are excluded to avoid absorbing states in the Markov chain. Moreover, one needs to exclude the alternating case where $p_{10} = 1 = p_{01}$. For simplicity, the cases $p_{01} = 1$ or $p_{10} = 1$ are also excluded to avoid deterministic transitions.

Hence, any transition matrix $P = (p_{ij})_{i,j \in \{0,1\}}$ considered in this thesis satisfies

$$p_{ij} > 0 \text{ for all } i,j \in \{0,1\}.$$

In particular, the Markov chain is aperiodic and irreducible and has a unique stationary distribution $\pi = \pi_0 \delta_0 + \pi_1 \delta_1$ given by

$$\pi_0 = \frac{p_{10}}{p_{10} + p_{01}}, \qquad \pi_1 = \frac{p_{01}}{p_{10} + p_{01}}. \tag{1.2}$$

Once again, a key parameter appearing in the analysis under the *Markov Source Model* is the *Source Entropy H*. Conditioned on the previous symbol in the string, there are two possible entropies compared to a *Bernoulli Source*

$$H_0 = -p_{00}\log(p_{00}) - p_{01}\log(p_{01}), \qquad H_1 = -p_{10}\log(p_{10}) - p_{11}\log(p_{11}). \tag{1.3}$$

The *Source Entropy* for the *Markov Source Model* is a weighted average of H_0 and H_1:

$$H := \pi_0 H_0 + \pi_1 H_1 \tag{1.4}$$

where π_0 and π_1 denote the weights of the stationary distribution given in (1.2) and H_0, H_1 are given in (1.3).

Note that the *Markov Source Model* covers the *Bernoulli Source Model* by choosing $p_{01} = p_{11} = p$ and $\mu = (1-p)\delta_0 + p\delta_1$. However, the results in section 1.4 do not cover the *symmetric Bernoulli Source Model* (i.e. $p_{ij} = 1/2$ for all $i,j \in \{0,1\}$). This is due to the fact that the asymptotic behavior of the variance of the structures in the symmetric model differs from the behavior in the asymmetric case.

The analysis of the *symmetric Bernoulli Source Model* can be found in the literature, cf. [17] for results on the mean of the depths in *Digital Trees*, [41] for the variance of the path length and [54] for a limit law derived with the *Contraction Method*. The connection between the path length in *Digital Trees* and the *Radix Sort* algorithm is explained in section 2.2.1.

1.3. DISTRIBUTIONAL RECURSIONS

The Density Model (Devroye 1982): The *Density Model* is an input model that is especially well suited when each string is considered to be the binary representation of a number in the unit interval $[0,1)$. Let f be a density function on $[0,1)$ and let X be a random variable with density f. The strings $\Xi_1, \ldots, \Xi_n$ are distributed according to the *Density Model* if they are independent and distributed as the binary expansion of X. More details on the *Density Model* can be found in [6, 7, 8].

Dynamical Sources (Vallée 2001): *Dynamical Sources* generalize the generator presented in the *Density Model*. As in the Density Model, let f be a density function on $[0,1)$ and let X be a random variable with density f. Moreover, let M be an arbitrary mechanism that associates a real number in $[0,1)$ with a (infinite) string, i.e.

$$M : [0,1) \longrightarrow \{(\xi_i)_{i \in \mathbb{N}} : \xi_i \in \Sigma \text{ for all } i \in \mathbb{N}\}.$$

Details on such mechanisms are given in [2].

In the *Dynamical Sources Model*, strings $\Xi_1, \ldots, \Xi_n$ are considered to be independent and distributed as $M(X)$.

This covers the *Density Model* by choosing $M(x)$ to be the binary expansion of the real number $x \in [0,1)$. It also covers the *Bernoulli* and *Markov Source Model* as explained in [2]. Mostly the first order asymptotic of the expectation of several parameters are derived in [2]. In particular, a limit law (and the asymptotic of the variance) for *Radix Sort* under the *Markov Source Model* was an open problem that is solved in this thesis.

1.3 Distributional Recursions

The recursive structure of *Radix Sort* leads to a distributional recursion for the number of *Bucket Operations* in the *Markov Source Model*. This section provides a description which explains how to derive such a distributional recursion and how this recursion needs to be generalized in order to cover similar recursive structures that appear in chapter 2.

Throughout this section, let $\Sigma = \{0, 1\}$ denote the binary alphabet and let $P = (p_{ij})_{i,j \in \Sigma}$ be a (fixed) stochastic matrix. For any initial distribution $\mu = \mu_0 \delta_0 + \mu_1 \delta_1$ and any integer n let $\mathcal{X}_n^\mu = [\Xi_1, \ldots, \Xi_n]$ be a list of n independent and identically distributed strings $\Xi_1, \ldots, \Xi_n$ where each string is generated by a *Markov Source* with initial distribution μ and transition matrix P (cf. definition 1.2.2).

Let B_n^μ denote the number of Bucket operations performed by Radix Sort with input $\mathcal{X}_n^\mu$. Moreover, let K_n^μ be the number of strings among $\mathcal{X}_n^\mu$ that start with the symbol 0.

Note that, conditioned on $K_n^\mu = k$, the algorithm performs n *Bucket Operations* in order to split $\mathcal{X}_n^\mu$ into a list $\mathcal{X}_k^{(0)}$ of k strings that start with the symbol 0 and a list $\mathcal{X}_{n-k}^{(1)}$ of $n-k$ strings that start with the symbol 1.

Also note that, by the Markov property, each string in $\mathcal{X}_k^{(0)}$ is a concatenation of the symbol 0 and a Markov chain with initial distribution $p_{00} \delta_0 + p_{01} \delta_1$ and transition matrix P (the independence assumption is also still valid for the strings in $\mathcal{X}_k^{(0)}$). Since the recursive call of *Radix Sort* in the sublist $\mathcal{X}_k^{(0)}$ does not consider the first symbol anymore, the number B_k^0 of *Bucket Operations* recursively sorting $\mathcal{X}_k^{(0)}$ is distributed as the number of *Bucket Operations* performed by *Radix*

Sort when sorting k independent and identically distributed strings generated by a *Markov Source* with initial distribution $p_{00}\delta_0 + p_{01}\delta_1$.

Similarly, the number of *Bucket Operations* recursively sorting the sublist $\mathcal{X}_{n-k}^{(1)}$ is distributed as the number of *Bucket Operations* performed by *Radix Sort* when sorting $n - k$ independent and identically distributed strings generated by a *Markov Source* with initial distribution $p_{10}\delta_0 + p_{11}\delta_1$.

Finally, due to the independence between the strings, the number of *Bucket Operations* performed in the two sublists are independent (conditioned on K_n^μ) and K_n^μ follows the binomial distribution $B(n, \mu_0)$.

This leads to the following distributional recursion for the number of *Bucket Operations*:

$$B_n^\mu \stackrel{d}{=} B_{K_n^\mu}^0 + B_{n-K_n^\mu}^1 + n, \quad n \geq 2, \tag{1.5}$$

with $(B_0^0, \ldots B_n^0)$, $(B_0^1, \ldots, B_n^1)$, K_n^μ independent and with distributions

$$\mathcal{L}(K_n^\mu) = B(n, \mu_0) \quad \text{and } \mathcal{L}\left(B_k^i\right) = \mathcal{L}\left(B_k^{p_{i0}\delta_0 + p_{i1}\delta_1}\right), \quad i \in \Sigma, \ k \geq 0.$$

In this context, the additive term n in (1.5) is sometimes called *toll term* because it covers the "cost" that is needed to split the problem of sorting the complete list into sorting two sublists. In other applications this cost may vary and does not need to be deterministic either.

The analysis of *PATRICIA Tries* in chapter 2 requires the consideration of more general *toll terms* than the constant n appearing in (1.5). Thus, this constant is replaced by a more general term η_n^μ which might depend on K_n^μ. More precisely, the *toll term* η_n^μ is considered to have the representation

$$\eta_n^\mu = g_n^\mu(K_n^\mu) \quad \text{for a (deterministic) function } g_n^\mu : \{0, \ldots, n\} \to \mathbb{R}. \tag{1.6}$$

Some additional requirements on g_n^μ are given in in section 1.4. In particular, the results in this thesis are restricted to linear growing *toll terms* in the sense that $\mathbb{E}[\eta_n^\mu] \sim n$ as $n \to \infty$.

In some other related recursive structures, such as *Digital Search Trees*, not all of the elements are distributed to the subproblems. In order to cover such problems as well, a generalized recursion of (1.5) needs to allow that only n out of $n + d$ elements are distributed to the subproblems for some fixed $d \in \mathbb{N}_0$.

Hence, the general framework of this thesis is to study a sequence $(X_n^\mu)_{n \geq 0}$ of random variables depending on a initial distribution μ that satisfy the distributional recursion

$$X_{n+d}^\mu \stackrel{d}{=} X_{K_n^\mu}^0 + X_{n-K_n^\mu}^1 + \eta_n^\mu \quad n \in \mathbb{N} \tag{1.7}$$

with $(X_0^0, \ldots, X_n^0)$, $(X_0^1, \ldots, X_n^1)$ and K_n^μ independent, $\mathcal{L}(K_n^\mu) = B(n, \mu_0)$ and

$$\mathcal{L}\left(X_k^i\right) = \mathcal{L}\left(X_k^{p_{i0}\delta_0 + p_{i1}\delta_1}\right) \quad \text{for } i \in \Sigma \text{ and } k \geq 0.$$

Here, $d \geq 0$ is a fixed integer and η_n^μ is a *toll term* with a representation given in (1.6).

The crucial part in the study of $(X_n^\mu)_{n \geq 0}$ is to handle the special cases $(X_n^0)_{n \geq 0}$ and $(X_n^1)_{n \geq 0}$ with initial distributions $p_{00}\delta_0 + p_{01}\delta_1$ and $p_{10}\delta_0 + p_{11}\delta_1$. The recursion (1.7) for these initial distributions becomes

$$\begin{aligned} X_{n+d}^0 &\stackrel{d}{=} X_{I_n^0}^0 + X_{n-I_n^0}^1 + \eta_n^0, \\ X_{n+d}^1 &\stackrel{d}{=} X_{I_n^1}^0 + X_{n-I_n^1}^1 + \eta_n^1, \end{aligned} \tag{1.8}$$

1.4. RESULTS ON RADIX SORT AND RELATED PROBLEMS

with $(X_0^0, \ldots, X_n^0)$, $(X_0^1, \ldots, X_n^1)$ and (I_n^0, I_n^1) independent, $\mathcal{L}(I_n^i) = B(n, p_{i0})$ and $\eta_n^i = g_n^i(I_n^i)$ for $i \in \Sigma$. The integer $d \geq 0$ and the functions $g_n^i := g_n^{p_{i0}\delta_0 + p_{i1}\delta_1}$ depend on the underlying recursive structure.

All results in the next section are derived by first studying X_n^0 and X_n^1 via (1.8). Afterwards, these results are transfered to arbitrary initial distributions via (1.7).

1.4 Results on Radix Sort and Related Problems

Before stating the main results concerning sequences $(X_n^\mu)_{n \geq 0}$ with distributional recursions (1.7), there are some additional assumptions needed concerning the integrability of X_n^μ and restrictions to the toll term η_n^μ.

Let $\Sigma = \{0, 1\}$ denote the binary alphabet and let $P = (p_{ij})_{i,j \in \Sigma}$ be a fixed transition matrix (stochastic matrix) that satisfies the conditions

$$p_{ij} \in (0,1) \text{ for all } (i,j) \in \Sigma^2, \qquad p_{ij} \neq \frac{1}{2} \text{ for some } (i,j) \in \Sigma^2. \tag{1.9}$$

Assume that $(X_n^\mu)_{n \geq 0}$ satisfies for any initial distribution μ the initial conditions

$$X_n^\mu = 0 \qquad \text{for all } n \leq \max\{d, 1\}, \tag{1.10}$$

where $d \geq 0$ is the fixed integer that appears in the distributional recursion (1.7).

Moreover, assume that there exists an $s \in (2, 3]$ such that

$$\mathbb{E}\left[|X_n^\mu|^s\right] < \infty \qquad \text{for all } n \in \mathbb{N}. \tag{1.11}$$

The restrictions on the toll term η_n^μ occurring in the recursion (1.7) are mainly needed for the special cases $\mu = p_{i0}\delta_0 + p_{i1}\delta_1$, $i \in \Sigma$, in order to analyze the system (1.8). The only assumptions needed for general μ in the transfer are that, as $n \to \infty$,

$$\mathbb{E}[\eta_n^\mu] = \mathrm{O}(n), \qquad \mathrm{Var}(\eta_n^\mu) = \mathrm{O}(n). \tag{1.12}$$

Additionally, assume that there exist constants $\varepsilon > 0$ and $C > 0$ such that the toll terms η_n^0 and η_n^1 appearing in the system (1.8) satisfy, as $n \to \infty$,

$$\begin{aligned}
\mathbb{E}[\eta_n^i] &= n + \mathrm{O}\left(n^{\frac{1}{2}-\varepsilon}\right), & \mathbb{E}[\Delta \eta_n^i] &= 1 + \mathrm{O}\left(n^{-\varepsilon}\right), \\
\mathrm{Var}(\eta_n^i) &= \mathrm{O}\left(n^{1-\varepsilon}\right), & \mathrm{Var}(\Delta \eta_n^i) &= \mathrm{O}(1), \\
\|\eta_n^i - \mathbb{E}[\eta_n^i]\|_3 &= o(\sqrt{n \log n}), & |\eta_n^i| &\leq Cn,
\end{aligned} \tag{1.13}$$

where $\Delta \eta_n^i := \eta_{n+1}^i - \eta_n^i$ denotes the (forward) difference of the sequence $(\eta_n^i)_{n \geq 0}$ for $i \in \Sigma$. To be precise about the embedding of η_{n+1}^i and η_n^i into a common probability space, recall that $\eta_n^i = g_n^i(I_n^i)$ for some function $g_n^i : \{0, \ldots, n\} \to \mathbb{R}$.

Therefore, let J_i be a random variable with Bernoulli distribution $B(p_{i0})$ that is independent of I_n^i and define

$$\Delta \eta_n^i \stackrel{d}{=} g_{n+1}^i(I_n^i + J_i) - g_n^i(I_n^i).$$

The following holds under these assumptions:

Theorem 1.4.1. *For arbitrary initial distributions μ let $(X_n^\mu)_{n\geq 0}$ be a sequence of real valued random variables that satisfies the stochastic recursion (1.7). Assume that the conditions (1.10)-(1.13) hold.*

Then, mean and variance of X_n^μ satisfy, as $n \to \infty$,

$$\mathbb{E}[X_n^\mu] = \frac{1}{H} n \log n + \mathrm{O}(n), \qquad \mathrm{Var}(X_n^\mu) = \sigma^2 n \log n + \mathrm{O}\left(n\sqrt{\log n}\right)$$

where the entropy rate H is defined in (1.4) and σ^2 is given by

$$\sigma^2 = \frac{\pi_0 p_{00} p_{01}}{H^3}\left(\log(p_{00}/p_{01}) + \frac{H_1 - H_0}{p_{01} + p_{10}}\right)^2 + \frac{\pi_1 p_{10} p_{11}}{H^3}\left(\log(p_{10}/p_{11}) + \frac{H_1 - H_0}{p_{01} + p_{10}}\right)^2.$$

Moreover, as $n \to \infty$,

$$\frac{X_n^\mu - \mathbb{E}[X_n^\mu]}{\sqrt{\mathrm{Var}(X_n^\mu)}} \xrightarrow{d} \mathcal{N}(0,1)$$

where $\mathcal{N}(0,1)$ denotes a random variable with the standard normal distribution.

The proof of theorem 1.4.1 is done in chapter 4 with techniques presented in chapter 3. Chapter 4 starts with the analysis of X_n^0 and X_n^1. In particular, the asymptotic analysis of the mean is done in section 4.1, the analysis of the variance is done in section 4.2 and the limit law is derived in section 4.3. These results are transferred to arbitrary initial distributions in section 4.4.

Note that the analysis of $\mathbb{E}[X_n^0]$ and $\mathbb{E}[X_n^1]$ also involves a more detailed study of the $\mathrm{O}(n)$ error term in their asymptotic representation. In fact, it is shown that $f_i(n) := \mathbb{E}[X_n^i] - \frac{1}{H} n \log n$ has bounded increments for both $i \in \Sigma$. This more detailed result is needed for the analysis of the variance and the limit law.

1.5 Results on Radix Select

The distributional equations discussed in the previous sections covered the *Radix Sort* algorithm. *Radix Select* is a one-sided version of this algorithm to select an element of a given rank:

Input: A list $\mathcal{X}_n = [\Xi_1, \ldots, \Xi_n]$ of n strings on an alphabet $\Sigma = \{0, \ldots, b-1\}$ and a rank $\ell \in \{1, \ldots, n\}$.

Output: The ℓ-th smallest element $\Xi_{(\ell)}$ among the strings in $\mathcal{X}_n$ (in lexicographical order).

The algorithm:

- Split $\mathcal{X}_n$ into b sublist $\mathcal{X}_{n,0}, \ldots, \mathcal{X}_{n,b-1}$ such that string Ξ_j is in list $\mathcal{X}_{n,i}$ if and only if the first symbol of Ξ_j is i, $j \in \{1, \ldots, n\}$, $i \in \Sigma$.

- Determine

$$m(\ell) := \min\left\{k \in \Sigma : \sum_{i=0}^{k} |\mathcal{X}_{n,i}| \geq \ell\right\}$$

where $|\mathcal{X}_{n,i}|$ denotes the number of strings in the i-th sublist. Note that the ℓ-th smallest element in $\mathcal{X}$ equals the $\ell - \sum_{i=0}^{m(\ell)-1} |\mathcal{X}_{n,i}|$ smallest element in the sublist $\mathcal{X}_{n,m(\ell)}$.

1.5. RESULTS ON RADIX SELECT

- If $|\mathcal{X}_{n,m(\ell)}| = 1$, return the string in $\mathcal{X}_{n,m(\ell)}$. Otherwise, search for the $\ell - \sum_{i=0}^{m(\ell)-1}|\mathcal{X}_{n,i}|$-th smallest element in $\mathcal{X}_{n,m(\ell)}$ by applying the *Radix Select* algorithm with a splitting according to the next symbol of each string in $\mathcal{X}_{n,m(\ell)}$.

Once again, it is quite natural to study the number of *Bucket Operations* performed by this algorithm where a *Bucket Operation* denotes the placement of a string into a sublist. Figure 1.2 shows an application of *Radix Select* on $n = 6$ strings.

Figure 1.2 The *Radix Select* algorithm on 6 strings searching for rank 2. Only the list that contains the element with rank 2 (green) is recursively split. The total number of bucket operations is $6 + 3 + 2 \cdot 2 = 13$.

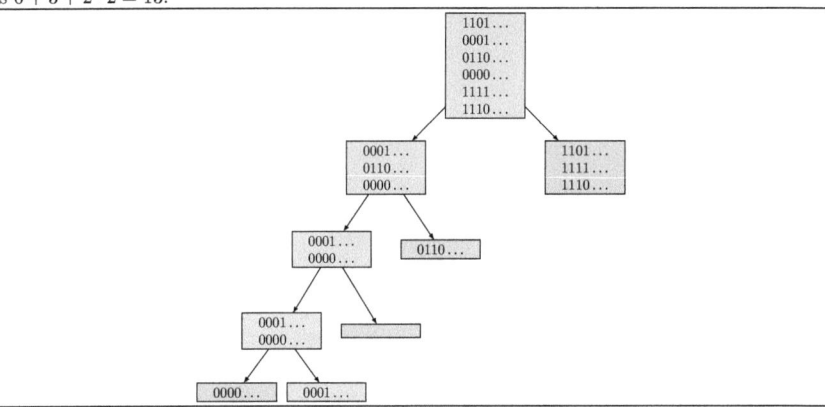

Input models: The list $\mathcal{X}_n$ may be modeled in several different ways. A few common stochastic models are summarized in section 1.2. The analysis in this thesis is focused on the *Markov Source Model* and a binary alphabet $\Sigma = \{0, 1\}$. Thus, it is assumed that $\mathcal{X}_n$ contains n independent and identically distributed strings $\Xi_1, \ldots, \Xi_n$ where the symbols of each string are distributed as a Markov chain on Σ. Details on the *Markov Source Model* are given in section 1.2.

Let $Y_n^\mu(\ell)$ denote the number of *Bucket Operations* performed by *Radix Select* when searching for the element of rank ℓ among n i.i.d. strings generated by a *Markov Source* with initial distribution $\mu = \mu_0 \delta_0 + \mu_1 \delta_1$ and transition matrix $P = (p_{ij})_{i,j \in \Sigma}$.

Moreover, let $Y_n^i(\ell) := Y_n^{p_{i0}\delta_0 + p_{i1}\delta_1}(\ell)$ for $n \in \mathbb{N}$ and $\ell \in \{1, \ldots, n\}$.

Throughout this section, it is assumed that the transition matrix satisfies

$$p_{ij} < 1 \text{ for all } i,j \in \Sigma.$$

The following result is shown in section 5.2.2:

Theorem 1.5.1. *The number of Bucket Operations performed by Radix Select when searching for an element of rank $\lfloor tn \rfloor + 1$ among n independent strings generated by a Markov Source with initial distribution $p_{i0}\delta_0 + p_{i1}\delta_1$ satisfies for all $t \in [0,1]$ and $i \in \Sigma$, as $n \to \infty$,*

$$\mathbb{E}[Y_n^i(\lfloor tn \rfloor + 1)] = m_i(t)n + o(n).$$

with $Y_n^i(n+1) := Y_n^i(n)$ and functions $m_i : [0,1] \to (0, \infty)$ defined in section 5.2.2.
In particular, m_i is a bounded function which is continuous on $[0,1] \setminus \mathcal{D}_\infty^i$. Here, $\mathcal{D}_\infty^i$ is a countable set determined by the transition matrix P. For any $t \in \mathcal{D}_\infty^i$, the limits $m_i(t-) := \lim_{s \uparrow t} m_i(s)$ and $m_i(t+) := \lim_{s \downarrow t} m_i(s)$ exist and the function satisfies $m_i(t) = (m_i(t-) + m_i(t+))/2$.
Moreover, the result also holds for an arbitrary initial distribution $\mu = \mu_0 \delta_0 + \mu_1 \delta_1$ and any $t \in [0,1] \setminus \mathcal{D}_\infty^\mu$ where $\mathcal{D}_\infty^\mu = (\mu_0 \mathcal{D}_\infty^0) \cup (\mu_1 \mathcal{D}_\infty^1 + \mu_0)$: in this case, the mean satisfies

$$\mathbb{E}[Y_n^\mu(\lfloor tn \rfloor + 1)] = m_\mu(t)n + o(n)$$

where the function $m_\mu : [0,1] \setminus \mathcal{D}_\infty^\mu \to \mathbb{R}$ is given by

$$m_\mu(t) = \begin{cases} \mu_0 m_0 \left(\frac{t}{\mu_0}\right) + 1, & \text{if } t < \mu_0, \\ (1-\mu_0) m_1 \left(\frac{t-\mu_0}{1-\mu_0}\right) + 1, & \text{if } t > \mu_0. \end{cases}$$

The theorem is shown by a suitable iteration of a distributional recursion of $(Y_n^i(\ell))_{\ell \in \{1,\ldots,n\}, n \in \mathbb{N}}$. The proof also requires some knowledge about the worst case behavior of *Radix Select*:

Theorem 1.5.2. *Let $M_n^i := \max_{\ell \in \{1,\ldots,n\}} Y_n^i(\ell)$, $i \in \Sigma$, denote the maximal number of Bucket Operations performed by Radix Select. Then, as $n \to \infty$,*

$$\frac{M_n^i}{n} \xrightarrow{d} \mathfrak{m}_i, \quad i \in \Sigma,$$

where $\mathfrak{m}_i = \sup_{t \in [0,1]} m_i(t)$ with $m_i : [0,1] \to \mathbb{R}$ given in theorem 1.5.1.

Moreover, for any $p > 0$,

$$\lim_{n \to \infty} \frac{1}{n^p} \mathbb{E}[(M_n^i)^p] = \mathfrak{m}_i^p.$$

In the special case $p_{00} = p_{10}$ (which corresponds to the *Bernoulli Source Model*), it is observed in [47] that

$$X_n^i := \left(\frac{Y_n^i(\lfloor tn \rfloor + 1) - m_i(t)n}{\sqrt{n}}\right)_{t \in [0,1]}, \quad n \in \mathbb{N},$$

converges in distribution (in the space of all càdlàg functions endowed with the Skorokhod topology) towards a centered Gaussian process. However, this does not hold for *Markov Sources* with $p_{00} \neq p_{10}$:

Theorem 1.5.3. *Consider a Markov Source with $p_{00} \neq p_{10}$. Then, for both $i \in \Sigma$, the family $\{\|X_n^i\|_\infty : n \in \mathbb{N}\}$ is not tight. In particular, the processes $(X_n^i)_{n \geq 0}$, $i \in \Sigma$, do not converge in distribution in $(\mathcal{D}[0,1], d_{sk})$. Here, $\mathcal{D}[0,1]$ denotes the space of all càdlàg functions $f : [0,1] \to \mathbb{R}$ and d_{sk} denotes the Skorokhod distance on $\mathcal{D}[0,1]$.*

In fact, the proof of the previous theorem also shows that any process

$$\left(\frac{Y_n^i(\lfloor tn \rfloor + 1) - m_i(t)n}{\alpha_n}\right)_{t \in [0,1]}$$

with $\alpha_n = o(n)$ cannot converge in distribution. A convergence of the marginals of X_n^i is not excluded by the previous theorem and remains an open problem not covered by this thesis.

1.5. RESULTS ON RADIX SELECT

Finally, there are some results on the *Grand Averages Model*: If the chosen rank is assumed to be uniformly distributed on $\{1,\ldots,n\}$ and independent of the strings $\Xi_1,\ldots,\Xi_n$, the following result holds for *Markov Sources*:

Theorem 1.5.4. *Let $W_n^i = Y_n^i(U_n)$ where U_n is uniformly distributed on $\{1,\ldots,n\}$ and independent of Y_n^i.*
Then, as $n \to \infty$,

$$\frac{W_n^i}{n} \xrightarrow{d} Z^i, \quad i \in \Sigma,$$

where $\mathcal{L}(Z^i) = \mathcal{L}(m_i(U))$ with the function m_i given in theorem 1.5.1 and a uniformly on $[0,1]$ distributed random variable U.

Moreover, the convergence also holds with all moments:

$$\frac{\mathbb{E}[(W_n^i)^p]}{n^p} \longrightarrow \mathbb{E}[(Z^i)^p], \quad i \in \Sigma, p > 0.$$

The limits Z^0 and Z^1 may be characterized as the unique pair of integrable solutions to a system (5.26) introduced in section 5.3. In particular, the expectations $\kappa_i := \mathbb{E}[Z^i]$, $i \in \Sigma$, are given by

$$\kappa_0 = \frac{1 + p_{01}^2 - p_{11}^2}{2(p_{00} + p_{11})(1 + p_{00}p_{11}) - 2(p_{00} + p_{11})^2},$$

$$\kappa_1 = \frac{1 + p_{10}^2 - p_{00}^2}{2(p_{00} + p_{11})(1 + p_{00}p_{11}) - 2(p_{00} + p_{11})^2}.$$

Most of the results on the *Grand Averages Model* are covered by a more detailed study of the process in theorem 1.5.1. However, all results on that model are (re-)derived in section 5.3 with the *Contraction Method* since this method provides an easy way to obtain the convergence of all moments combined with a characterization of the limiting distribution that allows to determine the moments of Z^i.

Chapter 2

Applications

There are several applications of theorem 1.4.1 in the field of random recursive structures and algorithms. In this chapter, the *Radix Sort* algorithm is studied in detail including the proof of the integrability condition (1.11).

Moreover, the asymptotic behavior of the path length in *Tries*, *PATRICIA Tries* and *Digital Search Trees* is discussed and analyzed by theorem 1.4.1. There is a generalization of *Digital Search Trees* where the capacity of each internal node in the tree is $b \geq 1$ including the standard Digital Search Tree with $b = 1$. Theorem 1.4.1 also applies to these structures by simply replacing the integer $d = 1$ in the analysis of the standard *Digital Search Tree* with $d = b$.

Some results on the related b-Tries under the *Bernoulli Source Model* can be found in [64] and the references therein.

Another application which, however, is not discussed in this thesis, is the analysis of the *Lempel-Ziv'78 Parsing Scheme* developed by the two authors in [69]. For the *Bernoulli Source Model* there is a well known transfer of results on the path length in *Digital Trees* to the asymptotic behavior of the number of blocks required by the *Lempel-Ziv'78 Parsing Scheme* to encode a random message of length n generated by a *Bernoulli Source*. An explanation of this connection and some results on other parameters in the *Markov Source Model* (excluding the number of blocks required) can be found in [37].

The transfer of the results concerning the path length in *Digital Search Trees* to the *Lempel-Ziv'78 Parsing Scheme* in the *Markov Source Model* turns out to be more difficult than in the *Bernoulli Source Model* due to the additional dependency between the parsed blocks. Therefore, the *Lempel-Ziv'78 Parsing Scheme* is not discussed at this point and details are left for further publications.

All applications are data structures and algorithms that rely on an input $\Xi_1, \ldots, \Xi_n$ of strings where each string is a sequence of symbols drawn from a (finite) alphabet Σ. Throughout this chapter, only the binary alphabet $\Sigma = \{0, 1\}$ is considered. Moreover, $\Xi_1, \ldots, \Xi_n$ are independent and distributed as a string Ξ that is generated by a *Markov Source* with an arbitrary initial distribution μ and a transition matrix $P = (p_{ij})_{i,j \in \Sigma}$ (see definition 1.2.2 on page 4 for details on *Markov Sources*).

The transition matrix P needs to satisfy the conditions (1.9), i.e.

$$p_{ij} \in (0,1) \text{ for all } (i,j) \in \Sigma^2, \qquad p_{ij} \neq \frac{1}{2} \text{ for some } (i,j) \in \Sigma^2.$$

2.1 Radix Sort

The first application is *Radix Sort*, the prime example given in chapter 1. Recall that B_n^μ denotes the number of *Bucket Operations* performed by *Radix Sort* applied to a list of n independent and identically distributed strings generated by a *Markov Source* with initial distribution μ and transition matrix P.

Theorem 1.4.1 yields the following result:

Corollary 2.1.1. *The number B_n^μ of Bucket operations performed by Radix Sort in the Markov Source Model satisfies for any initial distribution μ and any transition matrix P with conditions (1.9) that*

$$\mathbb{E}[B_n^\mu] = \frac{1}{H} n \log n + \mathrm{O}(n), \qquad \mathrm{Var}(B_n^\mu) = \sigma^2 n \log n + \mathrm{O}\left(n\sqrt{\log n}\right)$$

with constants H and σ^2 given in theorem 1.4.1. Moreover, as $n \to \infty$,

$$\frac{B_n^\mu - \mathbb{E}[B_n^\mu]}{\sqrt{\mathrm{Var}(B_n^\mu)}} \xrightarrow{d} \mathcal{N}(0,1)$$

where $\mathcal{N}(0,1)$ denotes a random variable with the standard normal distribution.

Proof. Let $\Xi_1, \ldots, \Xi_n$ denote the n independent and identically distributed input strings generated by a *Markov Source* with

$$\Xi_j = \left(\xi_k^{(j)}\right)_{k \geq 1}, \quad j = 1, \ldots, n.$$

As shown in section 1.3, B_n^μ satisfies the distributional recursion (1.5) which is

$$B_n^\mu \stackrel{d}{=} B_{K_n^\mu}^0 + B_{n-K_n^\mu}^1 + n, \qquad n \geq 2.$$

The only non-trivial condition that needs to be checked in order to apply theorem 1.4.1 is the integrability condition (1.11).

To this end, let $D_n^{(i)}$ be the number of *Bucket operations* involving Ξ_i. This leads to the decomposition

$$B_n^\mu = \sum_{i=1}^n D_n^{(i)}.$$

Note that $D_n^{(1)}, \ldots, D_n^{(n)}$ are identically distributed since the performance of *Radix Sort* is unaffected by any reordering of the strings $\Xi_1, \ldots, \Xi_n$.

Hence, the integrability condition (1.11) follows if, for some $s \in (2,3]$ and all $n \in \mathbb{N}$,

$$\mathbb{E}\left[\left|D_n^{(1)}\right|^s\right] < \infty.$$

2.2. DIGITAL TREES

In fact, this condition can be verified by the following equivalent description of $D_n^{(1)}$ (also motivated by the connection to *Tries* in section 2.2.1):

Note that $D_n^{(1)} \geq k$ if and only if there is at least one string with the same first $k-1$ symbols as Ξ_1 (such that the sublist containing Ξ_1 still needs to be sorted after $k-1$ iterations of the algorithm). Therefore, $D_n^{(1)}$ coincides with the length of the shortest prefix of Ξ_1 that does not appear in $\Xi_2, \ldots, \Xi_n$.

Also note that, for any $(s_1, \ldots, s_k) \in \Sigma^k$ and $k \geq 1$, by the i.i.d. assumption on $\Xi_2, \ldots, \Xi_n$

$$\mathbb{P}\left(\bigcap_{i=2}^n \left\{\left(\xi_1^{(i)}, \ldots, \xi_k^{(i)}\right) \neq (s_1, \ldots, s_k)\right\}\right) = \left(\mathbb{P}\left(\left(\xi_1^{(2)}, \ldots, \xi_k^{(2)}\right) \neq (s_1, \ldots, s_k)\right)\right)^{n-1}$$

$$= \left(1 - \mu_{s_1} \prod_{j=2}^k p_{s_{j-1} s_j}\right)^{n-1}$$

$$\geq \left(1 - p_\vee^{k-1}\right)^{n-1}$$

with $p_\vee = \max\{p_{ij} : i, j \in \Sigma\} < 1$. Hence, one obtains by conditioning on Ξ_1 that

$$\mathbb{P}\left(D_n^{(1)} \leq k\right) = \mathbb{P}\left(\bigcap_{i=2}^n \left\{\left(\xi_1^{(1)}, \ldots, \xi_k^{(1)}\right) \neq \left(\xi_1^{(j)}, \ldots, \xi_k^{(j)}\right)\right\}\right)$$

$$\geq \left(1 - p_\vee^{k-1}\right)^{n-1}$$

$$\geq 1 - (n-1) p_\vee^{k-1}$$

where the last bound holds by Bernoulli's inequality.

This yields

$$\mathbb{P}(D_n^{(1)} > k) \leq (n-1) p_\vee^{k-1}, \quad k \geq 1,$$

and therefore the finiteness of all moments of $D_n^{(1)}$ due to the exponential decay of the tails of the distribution.

This implies that B_n^μ has finite moments of any order. Hence, all conditions of theorem 1.4.1 are satisfied and the assertion follows. □

2.2 Digital Trees

The next application is the analysis of *Digital Trees*. The purpose of these trees is to store a finite set $\mathcal{X} = \{\Xi_1, \ldots, \Xi_n\}$ of strings (words) with symbols in some finite alphabet Σ. Although the analysis is only done when $\Sigma = \{0, 1\}$, the definition of *Digital Trees* is given with respect to an arbitrary alphabet $\Sigma = \{\sigma_1, \ldots, \sigma_m\}$, $m \geq 2$, .

There are three kinds of *Digital Trees* considered in this section: *Tries*, *PATRICIA Tries* and *Digital Search Trees*.

Tries: A *Trie* is an ordered, rooted tree structure that stores strings. The term *Trie* comes from (information) re*trie*val and is chosen due to the fact that *Tries* allow an efficient search for stored strings. The construction of a *Trie* is done recursively:

If the set $\mathcal{X}$ of strings is empty, then $Trie(\mathcal{X})$ is empty as well. If $|\mathcal{X}| = 1$, then $Trie(\mathcal{X})$ is a single node storing the string $\Xi_1 \in \mathcal{X}$. If $\mathcal{X}$ contains at least two strings, $\mathcal{X}$ is divided into $\mathcal{X}_a = \{(\xi_j)_{j \geq 2} : (a, \xi_2, \xi_3, \ldots) \in \mathcal{X}\}$, $a \in \Sigma$, and $Trie(\mathcal{X})$ is a tree that consists of a root node and subtrees $Trie(\mathcal{X}_{\sigma_1}), \ldots, Trie(\mathcal{X}_{\sigma_m})$.

Hence, in the special case $\Sigma = \{0, 1\}$ a *Trie* storing more than one string consists of a root and two subtrees, the left subtree storing every string that starts with symbol 0 and the right subtree storing every string that starts with symbol 1. The edges in a *Trie* refer to symbols in the alphabet. In particular, the edge connecting the left subtree to the root refers to the first symbol which is 0 for all strings stored in the left subtree.

Note that only the leaves of a *Trie* contain strings. These nodes are called *External Nodes*. All non-*External* nodes are called *Internal Nodes*. Figure 2.1 shows a *Trie* that stores the strings $\Xi_1, \ldots, \Xi_6$ with prefixes given by

$$\begin{aligned} \Xi_1 &= 1101\ldots, & \Xi_2 &= 0001\ldots, & \Xi_3 &= 0110\ldots, \\ \Xi_4 &= 0000\ldots, & \Xi_5 &= 1111\ldots, & \Xi_6 &= 1110\ldots \end{aligned} \qquad (2.1)$$

The black nodes in figure 2.1 are *Internal Nodes* and the blue nodes are *External Nodes*.

Figure 2.1 A *Trie* (a) and its reduction to a *PATRICIA Trie* (b). Both of them are storing the strings $\Xi_1, \ldots, \Xi_6$ listed in (2.1).

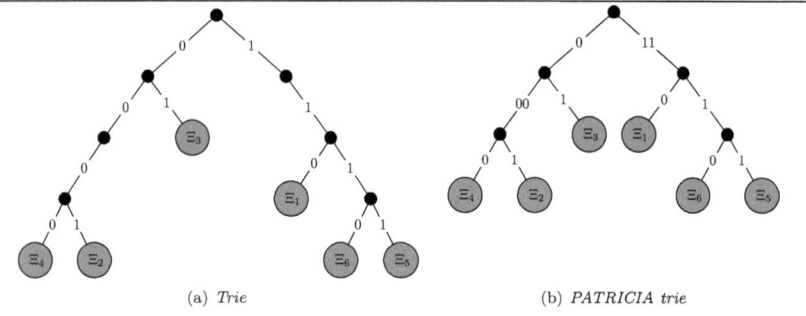

(a) *Trie*   (b) *PATRICIA trie*

PATRICIA Tries: A *PATRICIA Trie* is a compressed version of a *Trie* that eliminates nodes with a single branch. The term *PATRICIA* is an acronym for "Practical Algorithm To Retrieve Information Coded In Alphanumeric". The reduction from a *Trie* into a *PATRICIA Trie* works as follows: For each node u in the *Trie* that has exactly one child, merge u with its child, i.e. delete u and replace it by its child. Figure 2.1 shows the compressed version of the *Trie* containing the list (2.1).

There is also a direct recursive construction of *PATRICIA Tries*. For simplicity, only the case $\Sigma = \{0, 1\}$ is considered:

2.2. DIGITAL TREES

If the set $\mathcal{X}$ of strings is empty, then the *PATRICIA Trie* $PAT(\mathcal{X})$ is empty as well. If $\mathcal{X}$ contains a single string Ξ_1, then $PAT(\mathcal{X})$ is a single node storing Ξ_1. If $\mathcal{X}$ contains at least two strings, $\mathcal{X}$ is divided into $\mathcal{X}_0 = \{(\xi_j)_{j \geq 2} : (0, \xi_2, \xi_3, \ldots) \in \mathcal{X}\}$ and $\mathcal{X}_1 = \{(\xi_j)_{j \geq 2} : (1, \xi_2, \xi_3, \ldots) \in \mathcal{X}\}$ and there are three cases depending on the size of $\mathcal{X}_0$:

- If $\mathcal{X}_0$ is empty, let $PAT(\mathcal{X}) = PAT(\mathcal{X}_1)$;
- if $\mathcal{X}_1$ is empty, let $PAT(\mathcal{X}) = PAT(\mathcal{X}_0)$;
- otherwise, $PAT(\mathcal{X})$ consists of a root node with subtrees $PAT(\mathcal{X}_0)$ and $PAT(\mathcal{X}_1)$.

As in Tries, only the leaves of a *PATRICIA Trie* store strings. These leaves are called *External Nodes*.

Digital Search Trees: The strings in a *Digital Search Tree* are directly stored in the *Internal Nodes* of the tree. For simplicity, the construction of a *Digital Search Tree* is only described for the binary alphabet $\Sigma = \{0, 1\}$.

A *Digital Search Tree* that stores strings $\Xi_1, \ldots, \Xi_n$ is constructed as follows: the first string Ξ_1 is stored in the root of the *Digital Search Tree*. Then, Ξ_2 is stored in a node which is either the left or the right child of the root, depending on whether the first symbol of Ξ_2 is 0 or 1. The remaining strings traverse the tree according to their leading symbols (where 0 leads the string to the left child and 1 to the right child of a node) and are stored in the first empty position they visit. The maximal number of children per node increases for larger alphabets. An example for a *Digital Search Tree* that stores the strings (2.1) is given in figure 2.2.

Note that, other than in *Tries* and *PATRICIA Tries*, the shape of a *Digital Search Tree* depends on the ordering of $\Xi_1, \ldots, \Xi_n$.

Figure 2.2 A *Digital Search Tree* storing the strings listed in (2.1).

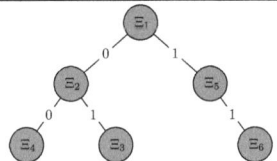

Parameters and Input Models: There are several parameters to study in *Digital Trees*. One of the most basic parameters is the depth $D_n^{(i)}$ of the node storing the i-th string in a *Digital Tree* holding n strings. Here, the depth of a node denotes its distance (number of edges) to the root. In the context of data structures, $D_n^{(i)}$ represents the searching cost when searching for string i.

The height H_n is another natural parameter when analyzing *Digital Trees*. It is given by

$$H_n = \max\left\{D_n^{(1)}, \ldots, D_n^{(n)}\right\}.$$

As the depths represent the searching costs, the height represents the worst case searching cost in a *Digital Tree*.

In addition to these two parameters, there is another depth-related parameter known as the (external) path length. The path length L_n of a *Digital Tree* is defined as

$$L_n = \sum_{i=1}^{n} D_n^{(i)}$$

and represents the construction cost of the *Digital Tree*. This parameter also encodes the average searching cost $\frac{1}{n}L_n$.

There are several other parameters not mentioned here, for example the profile and the size of *Digital Trees*. In fact, the distributional recursion (1.7) with a linear toll term is designed to cover the path length in *Digital Trees* and does not cover the other parameters.

Most of the studies of *Digital Trees* were done under the *Bernoulli Source Model*, see, e.g., [28, 16, 41, 42, 46, 51] for the examination of several parameters in *Tries*, *PATRICIA Tries* and *Digital Search Trees* under the *Bernoulli Source Model*. See also [19] for a general approach studying several parameters in *Digital Trees* with analytical methods and [54] for some general results derived by the *Contraction Method*.

Results for the height and depths of nodes in *Tries*, *PATRICIA Tries* and *Digital Search Trees* under the *Density Model* are given in [7, 8]. The external path length (or equivalently the average depth) of *Tries* under the *Density Model* is discussed in [6] including the first order asymptotic of the mean.

An asymptotic expansion for the mean of several parameters in *Tries* under the more general *Dynamical Sources Model* can be found in [2]. Moreover, there are laws of large numbers and concentration inequalities for the height and depth of *Tries* and *PATRICIA Tries* under a very general input model assumption given in [9, 10].

There are some results on the depth of *Tries* and *Digital Search Trees* for *Markov Sources* given in [31, 37] if the initial distribution of the source is the stationary distribution π given in (1.2) on page 4. Those results also cover asymptotic expansions for the mean of the path length in *Tries* and *Digital Search Trees* but they do not cover any results on the variance and limit laws for L_n.

For the remainder of this section, asymptotic results on mean and variance and a limit law is derived for the path length in *Tries*, *PATRICIA Tries* and *Digital Search Trees* under the *Markov Source Model*.

2.2.1 Analysis of Tries

The analysis of the external path length in *Tries* is covered by the result on *Radix Sort* given in section 2.1. This leads to the following result (which may also be shown by theorem 1.4.1):

Corollary 2.2.1. *Let TL_n^μ be the external path length of a Trie storing n independent and identically distributed strings where each string is generated by a Markov Source with an arbitrary initial distribution μ and a transition matrix P that satisfies the conditions (1.9).*

Then, mean and variance satisfy, as $n \to \infty$,

$$\mathbb{E}\left[TL_n^\mu\right] = \frac{1}{H} n \log n + \mathrm{O}(n), \qquad \mathrm{Var}\left(TL_n^\mu\right) = \sigma^2 n \log n + \mathrm{O}\left(n\sqrt{\log n}\right)$$

2.2. DIGITAL TREES

with constants H and σ^2 given in theorem 1.4.1. Moreover, as $n \to \infty$,

$$\frac{TL_n^\mu - \mathbb{E}[TL_n^\mu]}{\sqrt{\operatorname{Var}(TL_n^\mu)}} \xrightarrow{d} \mathcal{N}(0,1)$$

where $\mathcal{N}(0,1)$ denotes a random variable with the standard normal distribution.

Proof. Let $\mathcal{X}_n^\mu = [\Xi_1, \ldots, \Xi_n]$ be a list of n independent and identically distributed strings $\Xi_1, \ldots, \Xi_n$ where each string is generated by a *Markov Source* with initial distribution μ and transition matrix P.

Moreover, let $D_n^{(i)}$ denote the smallest integer k such that the k-prefix of string Ξ_i differs from the k-prefixes of each other string Ξ_j, $j \in \{1, \ldots, n\} \setminus \{i\}$. Here, the k-prefix of a string $\Xi = (\xi_\ell)_{\ell \geq 1}$ denotes the vector $(\xi_1, \ldots, \xi_k)$ of the first k-symbols of Ξ.

Finally, let TL_n^μ denote the external path length of a *Trie* storing the strings listed in $\mathcal{X}_n^\mu$ and let B_n^μ be the number of Bucket operations performed by Radix Sort with input $\mathcal{X}_n^\mu$.

The assertion then follows from corollary 2.1.1 and the equality

$$TL_n^\mu = \sum_{i=1}^n D_n^{(i)} = B_n^\mu.$$

This equality is quite obvious by the definition of *Radix Sort* and the *Trie*. It holds due to the fact that the splitting of the list into sublists in *Radix Sort* follows the same rules as the distribution of strings in *Tries* to the subtrees of the root which leads to the equality of the depth of Ξ_i in a *Trie* and the number of *Bucket Operations* involving Ξ_i. An illustration of the connection between *Tries* and *Radix Sort* is given in figure 2.3. □

Figure 2.3 *Radix Sort* and a *Trie* with input $\Xi_1, \ldots, \Xi_6$ given in (2.1). Note that Ξ_1 appears in four sublists (marked red) whereas the first three lead to a *Bucket Operation* involving Ξ_1. On the other hand, Ξ_1 has depth three in the corresponding *Trie*.

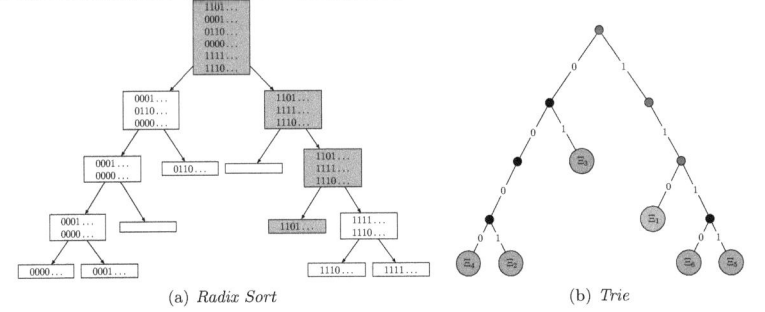

(a) *Radix Sort*   (b) *Trie*

2.2.2 Analysis of PATRICIA Tries

As a preparation for the analysis, the recursive description of *PATRICIA Tries* needs to be transformed into a distributional recursion of the external path length under the *Markov Source*

Model.

To this end, fix an initial distribution $\mu = \mu_0 \delta_0 + \mu_1 \delta_1$ and a transition matrix $P = (p_{ij})_{i,j \in \Sigma}$ and let $\mathcal{X} = \{\Xi_1, \ldots, \Xi_n\}$ be a set of n independent and identically distributed strings $\Xi_1, \ldots, \Xi_n$ where each string is generated by a *Markov Source* with initial distribution μ and transition matrix P.

Recall that $\mathcal{X}$ is split into the subsets

$$\mathcal{X}_i = \{(\xi_j)_{j \geq 2} : (i, \xi_2, \xi_3, \ldots) \in \mathcal{X}\}, \quad i \in \Sigma,$$

and $PAT(\mathcal{X})$ is constructed out of $PAT(\mathcal{X}_0)$ and $PAT(\mathcal{X}_1)$ according to the following rule that depends on the size K_n^μ of the subset $\mathcal{X}_0$:

(i) If $K_n^\mu = 0$, let $PAT(\mathcal{X}) = PAT(\mathcal{X}_1)$,

(ii) if $K_n^\mu = n$, let $PAT(\mathcal{X}) = PAT(\mathcal{X}_0)$,

(iii) otherwise, let $PAT(\mathcal{X})$ consist of a root node with the two subtrees $PAT(\mathcal{X}_0)$ and $PAT(\mathcal{X}_1)$.

Moreover, note that the independence and Markov property of the strings imply that, conditioned on $K_n^\mu = k$, $\mathcal{X}_0$ and $\mathcal{X}_1$ are two independent sets of strings where $\mathcal{X}_0$ holds k independent strings generated by a *Markov Source* with initial distribution $p_{00} \delta_0 + p_{01} \delta_1$ and $\mathcal{X}_1$ holds $n - k$ strings generated by a *Markov Source* with initial distribution $p_{10} \delta_0 + p_{11} \delta_1$. Both sources keep the original transition matrix P.

Also note that in the cases (i) and (ii), the path length of the complete *PATRICIA Trie* equals the path length of the non-empty subtree and in case (iii), the path length may be computed by deriving the path length of each subtree and adding the missing edge for each string. Therefore, the external path length PL_n^μ of a *PATRICIA Trie* with input $\mathcal{X}$ satisfies

$$PL_n^\mu \stackrel{d}{=} PL_{K_n^\mu}^0 + PL_{n-K_n^\mu}^1 + n \mathbf{1}_{\{K_n^\mu \notin \{0,n\}\}} \qquad (2.2)$$

with $(PL_k^0)_{k \geq 0}$, $(PL_k^1)_{k \geq 0}$ and K_n^μ independent and $\mathcal{L}(PL_k^i) = \mathcal{L}(PL_k^{p_{i0}\delta_0 + p_{i1}\delta_1})$ for $k \geq 0$ and $i \in \Sigma$. Also note that K_n^μ follows the binomial distribution $B(n, \mu_0)$ because all n strings are independent and the probability that a single string starts with symbol 0 equals μ_0.

Theorem 1.4.1 yields the following result for *PATRICIA Tries*:

Corollary 2.2.2. *Let PL_n^μ be the external path length of a PATRICIA Trie storing n independent and identically distributed strings where each string is generated by a Markov Source with an arbitrary initial distribution μ and a transition matrix P that satisfies the conditions (1.9).*

Then, mean and variance satisfy, as $n \to \infty$,

$$\mathbb{E}[PL_n^\mu] = \frac{1}{H} n \log n + O(n), \qquad \operatorname{Var}(PL_n^\mu) = \sigma^2 n \log n + O\left(n \sqrt{\log n}\right)$$

with constants H and σ^2 given in theorem 1.4.1. Moreover, as $n \to \infty$,

$$\frac{PL_n^\mu - \mathbb{E}[PL_n^\mu]}{\sqrt{\operatorname{Var}(PL_n^\mu)}} \stackrel{d}{\longrightarrow} \mathcal{N}(0,1)$$

where $\mathcal{N}(0,1)$ denotes a random variable with the standard normal distribution.

2.2. DIGITAL TREES

Proof. Recall that PL_n^μ satisfies the distributional recursion (2.2). Therefore, the only remaining task is to show that the conditions (1.10)-(1.13) of theorem 1.4.1 hold for PL_n^μ.

The initial condition (1.10) holds trivially for *PATRICIA Tries* (note that a *PATRICIA Trie* holding $n \leq 1$ strings is either empty or consists of a single root node).

The integrability condition (1.11) is implied by the integrability of the external path length in *Tries*. This is due to the fact that the *PATRICIA Trie* is a compressed version of a *Trie* and therefore,

$$PL_n^\mu \leq TL_n^\mu$$

where TL_n^μ denotes the external path length of a *Trie* with the same input as the *PATRICIA Trie*. The integrability of TL_n^μ (or equivalently the integrability of B_n^μ in *Radix Sort*) is discussed in the proof of corollary 2.1.1.

For the conditions on the toll term, note that K_n^μ follows the binomial distribution $B(n, \mu_0)$ which yields

$$\mathbb{E}\left[n\mathbf{1}_{\{K_n^\mu \notin \{0,n\}\}}\right] = n - n\left(\mu_0^n + \mu_1^n\right),$$
$$\mathrm{Var}\left(n\mathbf{1}_{\{K_n^\mu \notin \{0,n\}\}}\right) = n^2\left(\mu_0^n + \mu_1^n\right)\left(1 - \mu_0^n - \mu_1^n\right),$$
$$\left\|n\mathbf{1}_{\{K_n^\mu \notin \{0,n\}\}} - \mathbb{E}\left[n\mathbf{1}_{\{K_n^\mu \notin \{0,n\}\}}\right]\right\|_3 = n\left\|\mathbf{1}_{\{K_n^\mu \notin \{0,n\}\}} - (1 - \mu_0^n - \mu_1^n)\right\|_3$$
$$\leq n\left((\mu_0^n + \mu_1^n)^3 + (\mu_0^n + \mu_1^n)\right)^{1/3}.$$

Hence, the *toll term* satisfies for any $\mu = \mu_0 \delta_0 + \mu_1 \delta_1$ with $\mu_0 \in (0,1)$ that

$$\mathbb{E}\left[n\mathbf{1}_{\{K_n^\mu \notin \{0,n\}\}}\right] = n + o\left(n^{-1}\right),$$
$$\mathrm{Var}\left(n\mathbf{1}_{\{K_n^\mu \notin \{0,n\}\}}\right) = o\left(n^{-1}\right),$$
$$\left\|n\mathbf{1}_{\{K_n^\mu \notin \{0,n\}\}} - \mathbb{E}\left[n\mathbf{1}_{\{K_n^\mu \notin \{0,n\}\}}\right]\right\|_3 = o\left(n^{-1}\right)$$

and therefore, conditions (1.12) and (1.13) hold for all $\mu \notin \{\delta_0, \delta_1\}$. For the conditions on the differences note that

$$\mathbb{E}[\Delta \eta_m^\mu] = \mathbb{E}[\eta_{m+1}^\mu] - \mathbb{E}[\eta_m^\mu], \qquad \mathrm{Var}(\Delta \eta_n^\mu) = \mathrm{Var}(\eta_{n+1}^\mu) + \mathrm{Var}(\eta_n^\mu) + 2\mathrm{Cov}(\eta_{n+1}^\mu, \eta_n^\mu)$$

with $\mathrm{Cov}(\eta_{n+1}^\mu, \eta_n^\mu) \leq \sqrt{\mathrm{Var}(\eta_{n+1}^\mu)\mathrm{Var}(\eta_n^\mu)}$ by the Cauchy-Schwarz inequality.

Finally, note that the result also holds for $\mu \in \{\delta_0, \delta_1\}$ since in these cases the root always merges with its subtree and therefore

$$PL_n^{\delta_0} \stackrel{d}{=} PL_n^0, \qquad PL_n^{\delta_1} \stackrel{d}{=} PL_n^1.$$

□

2.2.3 Analysis of Digital Search Trees

The last application is the analysis of the path length in *Digital Search Trees*. The recursive behavior in *Digital Search Trees* is similar to the behavior in *Tries* with the slight difference that the first string is stored in the root instead of being distributed to the subtrees of the root.

Once again, let $\mu = \mu_0 \delta_0 + \mu_1 \delta_1$ be an arbitrary initial distribution and $P = (p_{ij})_{i,j\in\Sigma}$ be a transition matrix. Moreover, let $\mathcal{X} = [\Xi_1, \ldots, \Xi_{n+1}]$ be a list of $n+1$ independent and identically distributed strings generated by a *Markov Source* with initial distribution μ and transition matrix P.

Note that Ξ_1 is stored in the root of the *Digital Search Tree* and the remaining strings are distributed into the sublists

$$\mathcal{X}_i = [(\xi_j)_{j\geq 2} : (i, \xi_2, \xi_3, \ldots) \in \{\Xi_2, \ldots, \Xi_{n+1}\}], \quad i \in \Sigma.$$

The *Digital Search Tree* $DST(\mathcal{X})$ then consists of a root node with two subtrees $DST(\mathcal{X}_0)$ and $DST(\mathcal{X}_1)$ each of them following the same construction rule as the original *Digital Search Tree* with the changed input $\mathcal{X}_0$ and $\mathcal{X}_1$.

Using the same arguments as in the previous section, the path length DL^μ_{n+1} of a *Digital Search Tree* storing the strings listed in $\mathcal{X}$ satisfies

$$DL^\mu_{n+1} \stackrel{d}{=} DL^0_{K^\mu_n} + DL^1_{n-K^\mu_n} + n \quad (2.3)$$

with $(DL^0_k)_{k\geq 0}$, $(DL^1_k)_{k\geq 0}$ and K^μ_n independent, $\mathcal{L}(K^\mu_n) = B(n, \mu_0)$ and $\mathcal{L}(DL^i_k) = \mathcal{L}(DL^{p_{i0}\delta_0 + p_{i1}\delta_1}_k)$ for $k \geq 0$ and $i \in \Sigma$.

Theorem 1.4.1 yields:

Corollary 2.2.3. *Let DL^μ_n be the external path length of a Digital Search Tree storing n independent and identically distributed strings where each string is generated by a Markov Source with an arbitrary initial distribution μ and a transition matrix P that satisfies the conditions (1.9).*

Then, mean and variance satisfy, as $n \to \infty$,

$$\mathbb{E}[DL^\mu_n] = \frac{1}{H} n \log n + \mathrm{O}(n), \qquad \mathrm{Var}(DL^\mu_n) = \sigma^2 n \log n + \mathrm{O}\left(n\sqrt{\log n}\right)$$

with constants H and σ^2 given in theorem 1.4.1. Moreover, as $n \to \infty$,

$$\frac{DL^\mu_n - \mathbb{E}[DL^\mu_n]}{\sqrt{\mathrm{Var}(DL^\mu_n)}} \xrightarrow{d} \mathcal{N}(0, 1)$$

where $\mathcal{N}(0,1)$ denotes a random variable with the standard normal distribution.

Proof. Recall that DL^μ_n satisfies the distributional recursion (2.3). Hence, the only non trivial condition of theorem 1.4.1 that needs to be checked is the integrability condition (1.11).

However, the integrability of DL^μ_n is implied by the integrability of the external path length TL^μ_n of a *Trie* with the same input (it is also not hard to check that DL^μ_n is bounded by $n(n-1)/2$). More precisely,

$$DL^\mu_n \leq TL^\mu_n$$

because each string Ξ_i is stored at the position in a *Trie* that corresponds to the shortest prefix of Ξ_i which is unique among the prefixes of the input strings $\Xi_1, \ldots, \Xi_n$. This position cannot be occupied by $\Xi_1, \ldots, \Xi_{i-1}$ in a *Digital Search Tree* and therefore, Ξ_i is either stored at the same position in the *Digital Search Tree* or at another position which is closer to the root.

Hence, the analysis of *Tries* (or *Radix Sort*) implies that all moments of DL^μ_n are finite and theorem 1.4.1 yields the assertion. □

Chapter 3

Techniques

This chapter contains the tools required to derive the asymptotic of mean and variance as well as the limit theorems in chapter 4. The techniques introduced here focus on the analysis of the sequences $(X_n^0)_{n\geq 0}$ and $(X_n^1)_{n\geq 0}$. The transfer to arbitrary initial conditions does not need any particular preparation and is done in section 4.4.

Recall that $(X_n^0)_{n\geq 0}$ and $(X_n^1)_{n\geq 0}$ are considered to be sequences of real valued random variables that satisfy the initial conditions $X_n^0 = X_n^1 = 0$ for $n \leq \max\{d,1\}$ and the distributional recursion (1.8) given by

$$X_{n+d}^0 \stackrel{d}{=} X_{I_n^0}^0 + X_{n-I_n^0}^1 + \eta_n^0,$$
$$X_{n+d}^1 \stackrel{d}{=} X_{I_n^1}^0 + X_{n-I_n^1}^1 + \eta_n^1,$$ (3.1)

with $(X_0^0, \ldots, X_n^0)$, $(X_0^1, \ldots, X_n^1)$, (I_n^0, I_n^1) independent and $\mathcal{L}(I_n^i) = B(n, p_{i0})$, $i \in \Sigma$. Here, $\eta_n^i = g_n^i(I_n^i)$, $n \in \mathbb{N}_0$, $i \in \Sigma$, denotes a *toll term* that might depend on I_n^i.

Moreover, X_n^0 and X_n^1 are s-integrable for some $s \in (2,3]$ and constants $\varepsilon > 0$ and $C > 0$ exist in a way that the *toll terms* satisfy for both $i \in \Sigma$, as $n \to \infty$,

$$\mathbb{E}[\eta_n^i] = n + \mathrm{O}\left(n^{\frac{1}{2}-\varepsilon}\right), \qquad \mathbb{E}[\Delta\eta_n^i] = 1 + \mathrm{O}\left(n^{-\varepsilon}\right),$$
$$\mathrm{Var}(\eta_n^i) = \mathrm{O}\left(n^{1-\varepsilon}\right), \qquad \mathrm{Var}(\Delta\eta_n^i) = \mathrm{O}(1),$$
$$\|\eta_n^i - \mathbb{E}[\eta_n^i]\|_3 = o(\sqrt{n\log n}), \qquad |\eta_n^i| \leq Cn,$$

with $\Delta\eta_n^i := g_{n+1}^i(I_n^i + J_i) - g_n^i(I_n^i)$ for I_n^i and J_i independent, $\mathcal{L}(J_i) = B(p_{i0})$, $i \in \Sigma$.

Note that (3.1) implies recursive equations for mean and variance of X_n^i:

Lemma 3.0.4. *Let $\nu_i : \mathbb{N}_0 \to \mathbb{R}$ and $V_i : \mathbb{N}_0 \to \mathbb{R}_0^+$, $i \in \Sigma$, be defined as*

$$\nu_i(n) := \mathbb{E}[X_n^i], \qquad V_i(n) := \mathrm{Var}(X_n^i).$$

Then, (3.1) implies for all $n \in \mathbb{N}$ and $i \in \Sigma$

$$\nu_i(n+d) = \mathbb{E}[\nu_0(I_n^i)] + \mathbb{E}[\nu_1(n - I_n^i)] + \mathbb{E}[\eta_n^i] \qquad (3.2)$$
$$V_i(n+d) = \mathbb{E}[V_0(I_n^i)] + \mathbb{E}[V_1(n - I_n^i)] + \mathrm{Var}(\nu_0(I_n^i) + \nu_1(n - I_n^i) + \eta_n^i) \qquad (3.3)$$

Proof. Recall that in (3.1)

(a) $X^i_{n+d} \stackrel{d}{=} X^0_{I^i_n} + X^1_{n-I^i_n} + \eta^i_n$, $n \in \mathbb{N}$, $i \in \Sigma$,

(b) $(X^0_n)_{n \geq 0}$, $(X^1_n)_{n \geq 0}$ and $(I^0_n, I^1_n)_{n \geq 0}$ are independent.

Equation (3.2) is a direct consequence of taking the expectation in (a) and the fact that

$$\mathbb{E}\left[X^0_{I^i_n}\right] = \mathbb{E}\left[\mathbb{E}\left[X^0_{I^i_n} | I^i_n\right]\right] \stackrel{(b)}{=} \mathbb{E}\left[\nu_0(I^i_n)\right],$$
$$\mathbb{E}\left[X^1_{n-I^i_n}\right] = \mathbb{E}\left[\mathbb{E}\left[X^1_{n-I^i_n} | I^i_n\right]\right] \stackrel{(b)}{=} \mathbb{E}\left[\nu_1(n - I^i_n)\right],$$

$n \in \mathbb{N}$, $i \in \Sigma$.

For (3.3) note that for any $n \in \mathbb{N}$ and $i \in \Sigma$

$$V_i(n+d) \stackrel{(a)}{=} \mathbb{E}[(X^0_{I^i_n} + X^1_{n-I^i_n} + \eta^i_n - \mathbb{E}[X^i_{n+d}])^2]$$
$$= \mathbb{E}[(X^0_{I^i_n} - \nu_0(I^i_n) + X^1_{n-I^i_n} - \nu_1(n - I^i_n) + \nu_0(I^i_n) + \nu_1(n - I^i_n) + \eta^i_n - \mathbb{E}[X^i_{n+d}])^2]$$
$$\stackrel{(b)}{=} \mathbb{E}[V_0(I^i_n)] + \mathbb{E}[V_1(n - I^i_n)] + \mathbb{E}[(\nu_0(I^i_n) + \nu_1(n - I^i_n) + \eta^i_n - \mathbb{E}[X^i_{n+d}])^2]$$
$$\stackrel{(3.2)}{=} \mathbb{E}[V_0(I^i_n)] + \mathbb{E}[V_1(n - I^i_n)] + \operatorname{Var}(\nu_0(I^i_n) + \nu_1(n - I^i_n) + \eta^i_n)$$

which is the assertion. □

The asymptotic analysis of the mean requires a technique that transfers (3.2) and the asymptotic $\mathbb{E}[\eta^i_n] = n + \mathrm{O}\left(n^{1/2-\varepsilon}\right)$ into an asymptotic result for $\mathbb{E}[X^i_n]$. Such transfer lemmas (and similar lemmas for the study of the increments $(\Delta \nu_i(n))_{n \geq 0}$) are given in section 3.1.

The study of the variance is more involved because it takes a very detailed asymptotic expansion of ν_0 and ν_1 in order to derive the first order asymptotic of $\operatorname{Var}(\nu_0(I^i_n) + \nu_1(n - I^i_n) + \eta^i_n)$ in (3.3). Such an expansion seems far out of reach, at least with the methods presented in this thesis. Therefore, some additional methods are required for the variance.

In fact, by extending an idea in [63] it is possible to split X^i_n into a sum of two random variables $Y^i_n + Z^i_n$ such that an exact formula for the mean of $(Y^0_n)_{n \geq 0}$ and $(Y^1_n)_{n \geq 0}$ can be derived giving an opportunity to apply the transfer lemmas of section 3.1 to $(\operatorname{Var}(Y^0_n))_{n \geq 0}$ and $(\operatorname{Var}(Y^1_n))_{n \geq 0}$. Moreover, Z^i_n has an asymptotically negligible variance which is shown by another transfer result giving a connection between small toll terms (with small mean and variance) and a linear upper bound on the variance. This result relies on a *Poissonization* argument and some lemmas given in section 3.2.

The limit law for $(X^0_n)_{n \geq 0}$ and $(X^1_n)_{n \geq 0}$ is derived from an application of the *Contraction Method*. This method has successfully been applied to several parameters in the *Bernoulli Source Model*, see [54] for a general approach. An introduction to the *Contraction Method* is given in 3.3. There, the method is also extended to the *Markov Source Model*.

The adjustment to the *Markov Source Model* also includes a discussion on an appropriate study of the mean. In particular, section 3.3.3 gives an explanation why bounded increments in the error term $f_i(n) = \mathbb{E}[X^i_n] - \frac{1}{H} n \log n$ allow to apply the contraction method without a more detailed asymptotic expansion of the mean.

3.1 Asymptotic Analysis

Throughout this section, let $(a_0(n))_{n\in\mathbb{N}_0}$ and $(a_1(n))_{n\in\mathbb{N}_0}$ be two real valued sequences satisfying some sort of recursive equations. Typical examples are

$$a_i(n) = \mathbb{E}[X_n^i], \qquad a_i(n) = \text{Var}(X_n^i), \qquad a_i(n) = \Delta\mathbb{E}[X_n^i].$$

Moreover, let $(\varepsilon_0(n))_{n\in\mathbb{N}_0}$ and $(\varepsilon_1(n))_{n\in\mathbb{N}_0}$ be two real valued sequences that appear in the recursive equations of $(a_0(n))_{n\in\mathbb{N}_0}$ and $(a_1(n))_{n\in\mathbb{N}_0}$ as *toll terms*. *Toll terms* appearing in lemma 3.0.4 are

$$\varepsilon_i(n) = \mathbb{E}[\eta_n^i], \qquad \varepsilon_i(n) = \text{Var}(\nu_0(I_n^i) + \nu_1(n - I_n^i) + \eta_n^i), \qquad \varepsilon_i(n) = \mathbb{E}[\Delta \eta_n^i].$$

Finally, let $(I_n^0)_{n\in\mathbb{N}_0}$ and $(I_n^1)_{n\in\mathbb{N}_0}$ denote the binomial splitters in the recursive equations with distributions given by

$$\mathcal{L}(I_n^i) = B(n, p_i), \quad n \in \mathbb{N}_0, i \in \Sigma,$$

for some $p_0, p_1 \in (0, 1)$. In the study of a *Markov Source* with transition matrix $P = (p_{kl})_{k,l\in\Sigma}$, $p_i = p_{i0}$ for both $i \in \Sigma$.

The first lemma is a basic transfer between upper bounds on the *toll terms* and upper bounds $(a_0(n))_{n\in\mathbb{N}_0}$ and $(a_1(n))_{n\in\mathbb{N}_0}$:

Lemma 3.1.1. *Assume that there exists an integer $d \in \mathbb{N}_0$ such that for all $n \in \mathbb{N}$*

$$\begin{aligned} a_0(n + d) &= \mathbb{E}[a_0(I_n^0)] + \mathbb{E}[a_1(n - I_n^0)] + \varepsilon_0(n), \\ a_1(n + d) &= \mathbb{E}[a_0(I_n^1)] + \mathbb{E}[a_1(n - I_n^1)] + \varepsilon_1(n). \end{aligned} \quad (3.4)$$

If furthermore $\varepsilon_i(n) = O(n^\alpha)$ for an $\alpha \in \mathbb{R}$ and both $i \in \Sigma$, then, as $n \to \infty$,

$$a_i(n) = \begin{cases} O(n), & \text{if } \alpha < 1, \\ O(n^\alpha), & \text{if } \alpha > 1, \\ O(n \log n), & \text{if } \alpha = 1. \end{cases}$$

Proof. The proof relies on the fact that I_n^0 and I_n^1 are concentrated around their means $p_0 n$ and $p_1 n$. This leads to a geometric decay in the size of the toll term when iterating (3.4) on the right hand side.

It is more convenient to work with the monotone sequences given by

$$C_i(n) := \sup\{|a_i(k)| : 0 \le k \le n\}, \qquad C(n) := \max\{C_0(n), C_1(n)\}, \quad i \in \Sigma, n \in \mathbb{N}_0.$$

Due to the upper bound $|a_i(n)| \le C(n)$ for both $i \in \{0, 1\}$, an upper bound on $C(n)$ is sufficient to prove the assertion.

To this end, let $\delta \in (\max\{p_0, p_1, 1 - p_0, 1 - p_1\}, 1)$ be a constant (the exact value of δ does not matter) and decompose (3.4) into

$$\begin{aligned} a_i(n + d) &= \mathbb{E}[(a_0(I_n^i) + a_1(n - I_n^i))\mathbb{1}_{\{I_n^i \in [(1-\delta)n, \delta n]\}}] \\ &\quad + \mathbb{E}[(a_0(I_n^i) + a_1(n - I_n^i))\mathbb{1}_{\{I_n^i \notin [(1-\delta)n, \delta n]\}}] + \varepsilon_i(n). \end{aligned}$$

By assumption, there exists a constant $L > 0$ such that for all $n \in \mathbb{N}$ and $i \in \Sigma$

$$|\varepsilon_i(n)| \leq Ln^\alpha.$$

Applied to the decomposition above, the monotonicity of $C_0(n)$ and $C_1(n)$ yields

$$\begin{aligned}|a_i(n+d)| &\leq \mathbb{E}[(C_0(I_n^i) + C_1(n - I_n^i))\mathbb{1}_{\{I_n^i \in [(1-\delta)n, \delta n]\}}] \\ &\quad + (C_0(n+d) + C_1(n+d))\mathbb{P}(I_n^i \notin [(1-\delta)n, \delta n]) + Ln^\alpha \\ &\leq \mathbb{E}[(C(I_n^i) + C(n - I_n^i))\mathbb{1}_{\{I_n^i \in [(1-\delta)n, \delta n]\}}] \\ &\quad + 2C(n+d)\mathbb{P}(I_n^i \notin [(1-\delta)n, \delta n]) + Ln^\alpha\end{aligned}$$

Note that at least one of the following three equalities needs to hold by definition:

$$C(n) = |a_0(n)| \quad \text{or} \quad C(n) = |a_1(n)| \quad \text{or} \quad C(n) = C(n-1).$$

Therefore, for all $n \geq 1$ *at least one* of the following two bounds holds

$$\begin{aligned}\beta(n)C(n+d) &\leq \max_{i \in \Sigma}\left\{\mathbb{E}[(C(I_n^i) + C(n - I_n^i))\mathbb{1}_{\{I_n^i \in [(1-\delta)n, \delta n]\}}] + Ln^\alpha\right\} \\ C(n+d) &\leq C(n+d-1),\end{aligned} \qquad (3.5)$$

where $\beta(n) := 1 - 2\max_{i \in \Sigma}\{\mathbb{P}(I_n^i \notin [(1-\delta)n, \delta n])\}$ satisfies

$$\beta(n) \stackrel{n \to \infty}{\Longrightarrow} 1.$$

Now (3.5) and the upper bound $\mathbb{E}[(C(I_n^i) + C(n - I_n^i))\mathbb{1}_{\{I_n^i \in [(1-\delta)n, \delta n]\}}] \leq 2C(\lfloor \delta n \rfloor)$ imply for any $\varepsilon > 0$ by induction on n that

$$C(n) \leq Dn^{\max\{-\frac{\log 4}{\log \delta}, 2\alpha\}}(1+\varepsilon)^n$$

where $D = D(\varepsilon) > 0$ is a sufficiently large constant. This yields for any $K > 1$ the rough upper bound $C(n) = \mathrm{O}(K^n)$.

To refine this bound, note that by Chernoff's bound on the binomial distribution (see theorem A.1.1) there exists a constant $c > 0$ such that for all $n \geq 0$

$$|\beta(n) - 1| \leq 4e^{-cn}$$

which together with $C(n) = \mathrm{O}(K^n)$ for $1 < K < e^c$ implies that there exists a constant $L' > 0$ such that

$$|\beta(n) - 1|C(n+d) \leq L'n^\alpha, \quad n \in \mathbb{N}.$$

Now this bound and (3.5) yield that at least one of the following bounds holds

$$\begin{aligned}C(n+d) &\leq \max_{i \in \Sigma}\left\{\mathbb{E}[(C(I_n^i) + C(n - I_n^i))\mathbb{1}_{\{I_n^i \in [(1-\delta)n, \delta n]\}}] + (L+L')n^\alpha\right\} \\ C(n+d) &\leq C(n+d-1),\end{aligned}$$

which implies by induction on n that

$$C(n) \leq \tilde{L}n \sum_{j=0}^{\lfloor -\log n / \log \delta \rfloor} (\delta^{1-\alpha})^j$$

3.1. ASYMPTOTIC ANALYSIS

where $\tilde{L} = \max\{C(d+1), (L+L')\max\{\delta^{\alpha-1}, 1\}\}$. Finally, note that

$$\sum_{k=0}^{\lfloor -\log n/\log \delta \rfloor} \delta^{(1-\alpha)j} = \begin{cases} O(1), & \text{if } \alpha < 1, \\ O(\log n), & \text{if } \alpha = 1, \\ O(n^{\alpha-1}), & \text{if } \alpha > 1. \end{cases}$$

which yields the assertion. $\square$

The next lemma transfers the first order asymptotic of linear *toll terms* into a result on the first order asymptotic of $(a_0(n))_{n\geq 0}$ and $(a_1(n))_{n\geq 0}$. In particular, this lemma and lemma 3.0.4 yield the first order asymptotic of $\mathbb{E}[X_n^0]$ and $\mathbb{E}[X_n^1]$.

Lemma 3.1.2. *Assume that there exists an integer* $d \in \mathbb{N}_0$ *such that for all* $n \in \mathbb{N}$

$$\begin{aligned} a_0(n+d) &= \mathbb{E}[a_0(I_n^0)] + \mathbb{E}[a_1(n - I_n^0)] + \varepsilon_0(n), \\ a_1(n+d) &= \mathbb{E}[a_0(I_n^1)] + \mathbb{E}[a_1(n - I_n^1)] + \varepsilon_1(n). \end{aligned} \quad (3.6)$$

Moreover, assume that there exist constants $c_0, c_1 \in \mathbb{R}$ *and* $\alpha < 1$ *such that* $\varepsilon_i(n) = c_i n + O(n^\alpha)$ *for both* $i \in \Sigma$. *Then,* $a_i(n)$ *satisfies for both* $i \in \Sigma$, *as* $n \to \infty$,

$$a_i(n) = \frac{\pi_0 c_0 + \pi_1 c_1}{H} n \log n + O(n)$$

with constants π_0, π_1 *and* H *given by* $\pi_0 = \frac{p_1}{p_1 + 1 - p_0}, \pi_1 = \frac{1 - p_0}{p_1 + 1 - p_0}$ *and*

$$H = (-p_0 \log(p_0) - (1-p_0)\log(1-p_0))\pi_0 + (-p_1\log(p_1) - (1-p_1)\log(1-p_1))\pi_1.$$

Proof. Let $h : [0, \infty) \to \mathbb{R}$ be defined as

$$h(x) = \begin{cases} 0, & \text{if } x = 0, \\ x \log x, & \text{if } x > 0. \end{cases}$$

Moreover, let $H_i = -h(p_i) - h(1-p_i)$, $i \in \Sigma$, i.e. $H = \pi_0 H_0 + \pi_1 H_1$. Consider the transformed sequences $(\tilde{a}_0(n))_{n\geq 0}$ and $(\tilde{a}_1(n))_{n\geq 0}$ defined as

$$\tilde{a}_i(n) := a_i(n) - ch(n) + \frac{c_{1-i} H_i}{(p_1 + 1 - p_0)H} n$$

with $c = \frac{\pi_0 c_0 + \pi_1 c_1}{H}$. Note that the transformed sequences satisfy for all $n \in \mathbb{N}$ and $i \in \Sigma$

$$\tilde{a}_i(n+d) = \mathbb{E}[\tilde{a}_0(I_n^i)] + \mathbb{E}[\tilde{a}_1(n - I_n^i)] + \tilde{\varepsilon}_i(n)$$

with

$$\begin{aligned} \tilde{\varepsilon}_i(n) &= \varepsilon_i(n) - c\left(h(n+d) - \mathbb{E}[h(I_n^i) + h(n - I_n^i)]\right) \\ &\quad + \frac{c_{1-i}H_i}{(p_1+1-p_0)H}n - \frac{c_1 H_0}{(p_1+1-p_0)H}np_i - \frac{c_0 H_1}{(p_1+1-p_0)H}n(1-p_i). \end{aligned}$$

By lemma 3.1.1 and the definition of $(\tilde{a}_0(n))_{n\geq 0}$ and $(\tilde{a}_1(n))_{n\geq 0}$, it is sufficient to show

$$\tilde{\varepsilon}_i(n) = O\left(n^{\max\{\alpha, 1/3\}}\right).$$

To this end, note that

$$h(n+d) - \mathbb{E}[h(I_n^i) + h(n - I_n^i)]$$
$$= -\mathbb{E}[nh(I_n^i/n) + nh(1 - I_n^i/n)] + h(n+d) - h(n)$$
$$= H_i n - n\mathbb{E}[h(I_n^i/n) - h(p_i) + h(1 - I_n^i/n) - h(1 - p_i)] + h(n+d) - h(n)$$
$$= H_i n + h(n+d) - h(n) + \mathrm{O}\left(n^{1/3}\right)$$

where the last equality holds by the concentration of the binomial distribution and the asymptotic of $\log(1+x)$ as $x \to 0$ (note that $\log(I_n^i/n) - \log(p_i) = \log(1 + (I_n^i - np_i)/(np_i))$). Details can be found in lemma A.2.2 in the appendix. Finally,

$$h(n+d) - h(n) = d \log n + n \log\left(1 + \frac{d}{n}\right) = \mathrm{O}(\log n)$$

and therefore, by the assumption on $\varepsilon_i(n)$:

$$\tilde{\varepsilon}_i(n) = c_i n - c H_i n + \frac{c_{1-i} H_i}{(p_1 + 1 - p_0)H} n - \frac{c_1 H_0}{(p_1 + 1 - p_0)H} n p_0 + \mathrm{O}\left(n^{\max\{\alpha, 1/3\}}\right)$$
$$= n\left(c_i - \frac{p_1 c_0 + (1-p_0)c_1 - c_{1-i}}{(p_1 + 1 - p_0)H} H_i - \frac{p_i c_1 H_0 + (1-p_i)c_0 H_1}{(p_1 + 1 - p_0)H}\right) + \mathrm{O}\left(n^{\max\{\alpha, 1/3\}}\right)$$
$$= \mathrm{O}\left(n^{\max\{\alpha, 1/3\}}\right)$$

where the last equality can easily be seen by considering the cases $i = 0$ and $i = 1$ separately. The assertion follows by lemma 3.1.1 and the definition of $(\tilde{a}_0(n))_{n \geq 0}$ and $(\tilde{a}_1(n))_{n \geq 0}$. □

The next lemma is a useful tool in the analysis of the increments of sequences that satisfy (3.4). These increments satisfy similar recursive equations covered by the following transfer result:

Lemma 3.1.3. *Assume that there exist constants $d \in \mathbb{N}_0$ and $c_l \in (0,1)$, $l \in \Sigma$ such that for all $n \in \mathbb{N}$*

$$a_0(n+d) = c_0 \mathbb{E}[a_0(I_n^0)] + (1-c_0)\mathbb{E}[a_1(n - I_n^0)] + \varepsilon_0(n),$$
$$a_1(n+d) = c_1 \mathbb{E}[a_0(I_n^1)] + (1-c_1)\mathbb{E}[a_1(n - I_n^1)] + \varepsilon_1(n). \quad (3.7)$$

Then, as $n \to \infty$, $\varepsilon_0(n) = \mathrm{O}(n^\alpha)$ and $\varepsilon_1(n) = \mathrm{O}(n^\alpha)$ imply for both $i \in \Sigma$

$$a_i(n) = \begin{cases} \mathrm{O}(1) & \text{if } \alpha < 0, \\ \mathrm{O}(n^\alpha) & \text{if } \alpha > 0, \\ \mathrm{O}(\log n) & \text{if } \alpha = 0. \end{cases}$$

Proof. The idea is essentially the same as in the proof of lemma 3.1.1. However, there are a few changes due to the coefficients c_0 and c_1.

Once again, it is more convenient to work with the monotone sequences $(C_i(n))_{n \geq 0}$ and $(C(n))_{n \geq 0}$ given by

$$C_i(n) := \sup\{|a_i(k)| : 0 \leq k \leq n\}, \quad C(n) := \max\{C_0(n), C_1(n)\}, \quad n \in \mathbb{N}_0, i \in \Sigma.$$

Then, upper bounds on $C(n)$ also yield upper bounds on $|a_i(n)| \leq C(n)$ for both $i \in \Sigma$. Therefore, it only remains to derive upper bounds for $C(n)$.

3.1. ASYMPTOTIC ANALYSIS

To this end, let $\delta \in (\max\{p_0, p_1, 1 - p_0, 1 - p_1\}, 1)$ be a fixed constant and decompose (3.7) into

$$a_i(n+d) = \mathbb{E}[(c_i a_0(I_n^i) + (1-c_i)a_1(n - I_n^i))\mathbb{1}_{\{I_n^i \in [(1-\delta)n, \delta n]\}}]$$
$$+ \mathbb{E}[(c_i a_0(I_n^i) + (1-c_i)a_1(n - I_n^i))\mathbb{1}_{\{I_n^i \notin [(1-\delta)n, \delta n]\}}] + \varepsilon_i(n).$$

Together with the assumption $\varepsilon_i(n) = O(n^\alpha)$ and the monotonicity of $(C_0(n))_{n \geq 0}$ and $(C_1(n))_{n \geq 0}$, this implies for all $n \geq d+1$

$$|a_i(n)| \leq c_i C_i(\lfloor \delta n \rfloor) + (1-c_i)C_{1-i}(\lfloor \delta n \rfloor)$$
$$+ (c_i C_i(n) + (1-c_i)C_{1-i}(n))\mathbb{P}(I_{n-d}^i \notin [(1-\delta)(n-d), \delta(n-d)]) + Ln^\alpha \quad (3.8)$$

where $L > 0$ is some constant. The upper bound is derived in two steps: first, a rough upper bound $C(n) = O(K^n)$ for any $K > 1$ is derived; afterwards, the rough upper bound and Chernoff's bound on $\mathbb{P}(I_n^i \notin [(1-\delta)n, \delta n])$ are taken to derive a refined upper bound out of (3.8).

For the first step, note that (3.8) implies for all $n \geq d+1$

$$|a_i(n)| \leq C(\lfloor \delta n \rfloor) + C(n)\mathbb{P}(I_{n-d}^i \notin [(1-\delta)(n-d), \delta(n-d)]) + Ln^\alpha.$$

The definition of $C(n)$ yields for all $n \in \mathbb{N}$

$$C(n) = |a_0(n)| \quad \text{or} \quad C(n) = |a_1(n)| \quad \text{or} \quad C(n) = C(n-1)$$

and therefore,

$$\beta(n)C(n) \leq C(\lfloor \delta n \rfloor) + Ln^\alpha \quad \text{or} \quad C(n) = C(n-1) \quad (3.9)$$

where $\beta(n)$ is given by

$$\beta(n) = 1 - \max_{i \in \Sigma} \mathbb{P}(I_{n-d}^i \notin [(1-\delta)(n-d), \delta(n-d)]) \stackrel{n \to \infty}{\longrightarrow} 1.$$

Now, (3.9) implies for any $\varepsilon > 0$ by induction on $n \in \mathbb{N}$ that

$$C(n) \leq \widetilde{C} \cdot (1+\varepsilon)^{-\frac{\log n}{\log \delta}} \sum_{k=0}^{\lfloor -\log n / \log \delta \rfloor} \delta^{-\alpha k}$$

where $\widetilde{C} = \widetilde{C}(\varepsilon)$ is a sufficiently large constant. This yields a polynomial upper bound and, in particular,

$$C(n) = O(K^n) \quad \text{for all } K > 1.$$

In order to improve this bound, note that Chernoff's bound on $\mathbb{P}(I_n^i \notin [(1-\delta)n, \delta n])$ (see theorem A.1.1) implies that a constant $c > 0$ exists such that for all $n \in \mathbb{N}$, $i \in \Sigma$

$$\mathbb{P}(I_n^i \notin [(1-\delta)n, \delta n]) \leq 2e^{-cn}.$$

This fact, (3.8) and the rough upper bound with $1 < K < e^c$ yield that

$$|a_i(n)| \leq c_i C_i(\lfloor \delta n \rfloor) + (1-c_i)C_{1-i}(\lfloor \delta n \rfloor) + \hat{L}n^\alpha$$

where $\hat{L} > 0$ is a sufficiently large constant.

Hence, $C(n) = \max\{C_0(n), C_1(n)\}$ satisfies

$$C(n) \leq C(\lfloor \delta n \rfloor) + \hat{L}n^\alpha \quad \text{or} \quad C(n) = C(n-1)$$

which implies by induction on n that

$$C(n) \leq \tilde{L} \sum_{k=0}^{\lfloor -\log n/\log \delta \rfloor} \delta^{-\alpha j}$$

where $\tilde{L} = \max\{C(d+1), \max\{\delta^{-\alpha}, 1\}\hat{L}\}$.

Finally, the asymptotic behavior of the sum implies the assertion. $\square$

There is also a transfer result on bounded *toll terms* with given first order asymptotic:

Lemma 3.1.4. *Assume that there exist constants $d \in \mathbb{N}_0$ and $c_l \in (0,1)$, $l \in \Sigma$ such that for all $n \in \mathbb{N}$*

$$\begin{aligned} a_0(n+d) &= c_0 \mathbb{E}[a_0(I_n^0)] + (1-c_0)\mathbb{E}[a_1(n-I_n^0)] + \varepsilon_0(n), \\ a_1(n+d) &= c_1 \mathbb{E}[a_0(I_n^1)] + (1-c_1)\mathbb{E}[a_1(n-I_n^1)] + \varepsilon_1(n). \end{aligned} \qquad (3.10)$$

Moreover, assume that $\varepsilon_0(n) = \varepsilon_0 + \mathrm{O}(n^{-\alpha})$ and $\varepsilon_1(n) = \varepsilon_1 + \mathrm{O}(n^{-\alpha})$ for some constants $\varepsilon_0, \varepsilon_1 \in \mathbb{R}$ and $\alpha > 0$.

Then, as $n \to \infty$,

$$a_i(n) = L \log n + \mathrm{O}(1), \quad i \in \Sigma,$$

with a constant $L \in \mathbb{R}$ given by

$$L = \frac{c_1 \varepsilon_0 + (1-c_0)\varepsilon_1}{-c_1(c_0 \log p_0 + (1-c_0)\log(1-p_0)) - (1-c_0)(c_1 \log p_1 + (1-c_1)\log(1-p_1))}.$$

Remark 3.1.5. *In particular, lemma 3.1.4 yields for constants $c_0, c_1, \varepsilon_0, \varepsilon_1$ with*

$$c_1 \varepsilon_0 + (1-c_0)\varepsilon_1 = 0$$

that $a_i(n) = \mathrm{O}(1)$ for both $i \in \Sigma$.

Proof of lemma 3.1.4. Let $g : \mathbb{N}_0 \to \mathbb{R}$ be defined as

$$g(x) = \log(x+1).$$

Consider the rescaled sequences $(\tilde{a}_0(n))_{n \in \mathbb{N}_0}$ and $(\tilde{a}_1(n))_{n \in \mathbb{N}_0}$ given by

$$\begin{aligned} \tilde{a}_0(n) &:= a_0(n) - Lg(n) - \frac{\varepsilon_1(c_0 \log p_0 + (1-c_0)\log(1-p_0))}{C}, \\ \tilde{a}_1(n) &:= a_1(n) - Lg(n) - \frac{\varepsilon_0(c_1 \log p_1 + (1-c_1)\log(1-p_1))}{C} \end{aligned}$$

with $C = -c_1(c_0 \log p_0 + (1-c_0)\log(1-p_0)) - (1-c_0)(c_1 \log p_1 + (1-c_1)\log(1-p_1))$.

Then, (3.10) implies

$$\tilde{a}_0(n) = c_0 \mathbb{E}[\tilde{a}_0(I_n^0)] + (1-c_0)\mathbb{E}[\tilde{a}_1(n-I_n^0)] + \tilde{\varepsilon}_0(n)$$

with a *toll term* given by

$$\begin{aligned} \tilde{\varepsilon}_0(n) = \varepsilon_0(n) - Lg(n) + c_0 L \mathbb{E}[g(I_n^0)] + (1-c_0) L \mathbb{E}[g(n-I_n^0)] \\ + (1-c_0) \frac{\varepsilon_0(c_1 \log p_1 + (1-c_1)\log(1-p_1)) - \varepsilon_1(c_0 \log p_0 + (1-c_0)\log(1-p_0))}{C}. \end{aligned}$$

3.1. ASYMPTOTIC ANALYSIS

Note that Chernoff's bound on the binomial distribution implies $\mathbb{E}[g(I_n^0)] = g(n) + \log(p_0) + \mathrm{O}\left(n^{-1/2}\right)$ and $\mathbb{E}[g(n - I_n^0)] = g(n) + \log(1 - p_0) + \mathrm{O}\left(n^{-1/2}\right)$ (details given in lemma A.2.2). Hence, the assumption on $\varepsilon_0(n)$ yields

$$\tilde{\varepsilon}_0(n) = \frac{C\varepsilon_0 + (c_1\varepsilon_0 + (1-c_0)\varepsilon_1)\left(c_0 \log p_0 + (1-c_0) \log(1-p_0)\right)}{C}$$
$$+ (1-c_0)\frac{\varepsilon_0\left(c_1 \log p_1 + (1-c_1) \log(1-p_1)\right) - \varepsilon_1\left(c_0 \log p_0 + (1-c_0) \log(1-p_0)\right)}{C}$$
$$+ \mathrm{O}\left(n^{-\min\{\alpha, 1/2\}}\right)$$
$$= \mathrm{O}\left(n^{-\min\{\alpha, 1/2\}}\right).$$

A similar calculation reveals that

$$\tilde{a}_1(n) = c_1 \mathbb{E}[\tilde{a}_0(I_n^1)] + (1-c_1)\mathbb{E}[\tilde{a}_1(n - I_n^1)] + \mathrm{O}\left(n^{-\min\{\alpha, 1/2\}}\right).$$

Hence, lemma 3.1.3 yields that $(\tilde{a}_0(n))_{n \in \mathbb{N}_0}$ and $(\tilde{a}_1(n))_{n \in \mathbb{N}_0}$ are bounded and the assertion follows. $\square$

The last transfer result in this section is dedicated to the analysis of Radix Select in chapter 5.

Lemma 3.1.6. *Let $(a(n))_{n \in \mathbb{N}_0}$ be a real valued sequence and I_n a binomial $B(n, p)$ distributed random variable for some $p \in (0, 1)$. Suppose that there exists a real valued sequence $(\varepsilon(n))_{n \in \mathbb{N}}$ such that for all $n \in \mathbb{N}$*

$$a(n) = \mathbb{E}[a(I_n)] + \varepsilon(n).$$

Then, as $n \to \infty$, $\varepsilon(n) - \mathrm{O}(n^\alpha)$ for some $\alpha \in \mathbb{R}$ implies

$$a(n) = \begin{cases} \mathrm{O}(1), & \text{if } \alpha < 0, \\ \mathrm{O}(\log n), & \text{if } \alpha = 0, \\ \mathrm{O}(n^\alpha), & \text{if } \alpha > 0. \end{cases}$$

Proof. Let $C(n) := \sup\{|a(k)| : 0 \leq k \leq n\}$ and $\delta \in (p, 1)$. As in the previous proof, note that

$$C(n) \leq C(n-1) \quad \text{or} \quad \beta(n)C(n) \leq C(\lfloor \delta n \rfloor) + Ln^\alpha$$

with $\beta(n) = 1 - \mathbb{P}(I_n > \delta n) \to 1$.

Note that this bound also appeared in (3.9). As it is shown there, one obtains the upper bound

$$C(n) = \mathrm{O}(K^n) \quad \text{for all } K > 1.$$

Moreover, the arguments in the proof of lemma 3.1.3 also yield the refined upper bound

$$C(n) \leq \tilde{L} \sum_{k=0}^{\lfloor -\log n / \log \delta \rfloor} \delta^{-\alpha j}$$

for some suitable constant $\tilde{L} > 0$ and the assertion follows. $\square$

3.2 Poissonization and Depoissonization

A very useful tool in the asymptotic analysis of variances is *Poissonization*. The idea of *Poissonization* is to replace the fixed number n with a Poisson $\Pi(\lambda)$ distributed random variable N_λ and derive the (first order) asymptotic of $\mathrm{Var}(X_{N_\lambda}^i)$ as $\lambda \to \infty$.

This turns out to be easier than analyzing the stochastic recursion (3.1), at least for toll terms with small mean and variance.

More precisely, for any $\lambda > 0$, let N_λ be a Poisson $\Pi(\lambda)$ distributed random variable that is independent of $\{X_n^0, X_n^1, I_n^0, I_n^1 : n \in \mathbb{N}_0\}$. Then, (3.1) implies for both $i \in \Sigma$ and any $\lambda > 0$

$$X_{N_\lambda+d}^i \stackrel{d}{=} X_{N_{\lambda p_{i0}}}^0 + X_{M_{\lambda p_{i1}}}^1 + \eta_{N_\lambda}^i \qquad (3.11)$$

where $N_{\lambda p_{i0}} := I_{N_\lambda,0}^i$, $M_{\lambda p_{i1}} := I_{N_\lambda,1}^i$ and $(X_n^0)_{n\geq 0}$, $(X_n^1)_{n\geq 0}$, $(I_n^i)_{n\geq 0}$ and N_λ are independent.

Note that, as a well known fact from *Poisson Processes* (marking each point with probability p_{i0}), $N_{\lambda p_{i0}}$ and $M_{\lambda p_{i1}}$ are *independent* and Poisson distributed. Hence, $X_{N_{\lambda p_{i0}}}^0$ and $X_{M_{\lambda p_{i1}}}^1$ are independent even without conditioning on $N_{\lambda p_{i0}}$ and $M_{\lambda p_{i1}}$.

The transfer result in this section is done for functions

$$h_i : \mathbb{R}^+ \to \mathbb{R}, \qquad \eta_i : \mathbb{R}^+ \to \mathbb{R}, \quad i \in \Sigma$$

that satisfy for all $x \in \mathbb{R}^+$

$$\begin{aligned} h_0(x) &= h_0(xp_{00}) + h_1(xp_{01}) + \eta_0(x), \\ h_i(x) &= h_0(xp_{10}) + h_1(xp_{11}) + \eta_i(x) \end{aligned} \qquad (3.12)$$

with constants $p_{i0} \in (0,1)$ and $p_{i1} = 1 - p_{i0}$ for both $i \in \Sigma$.

Indeed, $h_i(x) = \mathrm{Var}(X_{N_x}^i)$, $i \in \Sigma$, satisfies (3.12) for sequences $(X_n^0)_{n\geq 0}$ and $(X_n^1)_{n\geq 0}$ such that (3.11) holds with $d = 0$. For $d \geq 1$, one additionally needs upper bounds on the difference $|\mathrm{Var}(X_{N_\lambda+d}^i) - \mathrm{Var}(X_{N_\lambda}^i)|$. Such bounds can be derived by the conditioning on N_λ and the following lemma:

Lemma 3.2.1. *Let $(a(n))_{n\in\mathbb{N}_0}$ be a real valued sequence and $d \in \mathbb{N}$ be some constant. Moreover, let N_λ be Poisson(λ) distributed for $\lambda > 0$.*

Then, $a(n) = \mathrm{O}(n\log n)$ implies for all $\varepsilon > 0$, as $\lambda \to \infty$,

$$\mathbb{E}[a(N_\lambda + d)] = \mathbb{E}[a(N_\lambda)] + \mathrm{O}\left(\lambda^{\frac{2}{3}+\varepsilon}\right).$$

Proof. First note that $a(n) = \mathrm{O}(n\log n)$ implies that a constant $C > 0$ exists such that

$$|\mathbb{E}[a(N_\lambda + d - 1)]| \leq C\mathbb{E}[(N_\lambda + d - 1)\log(N_\lambda + d - 1)] = \mathrm{O}(\lambda \log \lambda) \qquad (3.13)$$

where the last equality is easy to check (see appendix, lemma A.2.3). Moreover, an easy calculation (also known from Stein's method) reveals

$$\mathbb{E}[a(N_\lambda + d)] = \frac{1}{\lambda}\mathbb{E}[N_\lambda a(N_\lambda + d - 1)].$$

3.2. POISSONIZATION AND DEPOISSONIZATION

The assertion follows by induction on d and the following bound:

$$\left|\frac{1}{\lambda}\mathbb{E}[N_\lambda a(N_\lambda + d - 1)] - \mathbb{E}[a(N_\lambda + d - 1)]\right| = O\left(\lambda^{\frac{2}{3}+\varepsilon}\right). \tag{3.14}$$

Therefore, it is sufficient to show that

$$\mathbb{E}[N_\lambda a(N_\lambda + d - 1)] \le \left(\lambda + \lambda^{\frac{2}{3}}\right)\mathbb{E}[a(N_\lambda + d - 1)] + O\left(\lambda^{\frac{5}{3}+\varepsilon}\right), \tag{3.15}$$

$$\mathbb{E}[N_\lambda a(N_\lambda + d - 1)] \ge \left(\lambda - \lambda^{\frac{2}{3}}\right)\mathbb{E}[a(N_\lambda + d - 1)] + O\left(\lambda^{\frac{5}{3}+\varepsilon}\right), \tag{3.16}$$

which combined with (3.13) yields the desired bound (3.14).

Both bounds can be computed by the following decompositions

$$\mathbb{E}[N_\lambda a(N_\lambda + d - 1)] \le \mathbb{E}[(\lambda + \lambda^{2/3})a(N_\lambda + d - 1)] + \mathbb{E}[N_\lambda a(N_\lambda + d - 1)\mathbb{1}_{\{N_\lambda > \lambda + \lambda^{2/3}\}}],$$

$$\mathbb{E}[N_\lambda a(N_\lambda + d - 1)] \ge \mathbb{E}[(\lambda - \lambda^{2/3})a(N_\lambda + d - 1)] - \mathbb{E}[(\lambda - \lambda^{2/3})a(N_\lambda + d - 1)\mathbb{1}_{\{N_\lambda < \lambda - \lambda^{2/3}\}}].$$

The bounds on the remaining terms

$$\mathbb{E}[N_\lambda a(N_\lambda + d - 1)\mathbb{1}_{\{N_\lambda > \lambda + \lambda^{2/3}\}}], \qquad \mathbb{E}[(\lambda - \lambda^{2/3})a(N_\lambda + d - 1)\mathbb{1}_{\{N_\lambda < \lambda - \lambda^{2/3}\}}]$$

follow from Hölder's inequality, an upper bound on $\mathbb{P}(N_\lambda > \lambda + \lambda^{2/3})$ and $\mathbb{P}(N_\lambda < \lambda - \lambda^{2/3})$ given by Chebyshev's inequality, the asymptotic of $a(n)$ and the fact that, as $\lambda \to \infty$,

$$\mathbb{E}\left[N_\lambda^\alpha (\log N_\lambda)^\beta\right] = O\left(\lambda^\alpha (\log \lambda)^\beta\right) \quad \text{for all } \alpha, \beta > 0.$$

Details on the asymptotic behavior of $\mathbb{E}[N_\lambda^\alpha (\log N_\lambda)^\beta]$ are given in lemma A.2.3 on page 115. □

The next lemma provides a way to deduce the asymptotic behavior of functions h_i that satisfy (3.12).

Lemma 3.2.2. Let h_0 and h_1 be some functions that satisfy (3.12) and $\sup_{x \le a} |h_i(x)| < \infty$ for all $i \in \Sigma$ and $a \in \mathbb{R}^+$.

Then, as $x \to \infty$, $\eta_i(x) = O(x^{1-\alpha})$ for some $\alpha > 0$ and both $i \in \Sigma$ implies

$$h_i(x) = O(x), \quad i \in \Sigma.$$

Proof. Let $p \in (0,1)$ and $n_0 \in \mathbb{N}$ be defined as

$$p := \max\{p_{ij} : i, j \in \Sigma\}, \qquad n_0 := \lceil \log(\min\{p_{i,j} : i, j \in \Sigma\})/\log p \rceil,$$

i.e. $p_{ij} x \ge 1$ for all $x \ge p^{-n_0}$.

The assertion follows if the following holds for some constant $C > 0$ and all $n \in \mathbb{N}$:

$$|h_0(x)| \le Cx \sum_{j=0}^{\lfloor -\frac{\log x}{\log p} \rfloor} p^{\alpha j}, \quad x \in [1, p^{-n}],$$

$$|h_1(x)| \le Cx \sum_{j=0}^{\lfloor -\frac{\log x}{\log p} \rfloor} p^{\alpha j}, \quad x \in [1, p^{-n}].$$

(3.17)

The proof of (3.17) is done by induction on n.

Note that, by assumption, there exists a constant $\widetilde{C} > 0$ and an integer $n_1 \geq n_0$ such that $|\eta_i(x)| \leq \widetilde{C}x^{1-\alpha}$ for all $x > p^{-n_1}$ and $i \in \Sigma$. In particular,

$$|h_i(x)| \leq |h_0(xp_{i0})| + |h_1(xp_{i1})| + \widetilde{C}x^{1-\alpha} \quad \text{for all } x > p^{-n_1} \text{ and } i \in \Sigma. \tag{3.18}$$

Now let
$$C := \max\left\{\widetilde{C}, \max\{|h_i(x)| : i,j \in \Sigma, x \in (0, p^{-n_1}]\}\right\} < \infty.$$

Then, (3.17) holds trivially for $n \leq n_1$.

For $n > n_1$ and $x \in (p^{-(n-1)}, p^{-n}]$, note that $p_{ij}x \in [1, p^{-(n-1)}]$ for all $i, j \in \Sigma$ and therefore, by (3.18) and the induction hypothesis

$$\begin{aligned}
|h_i(x)| &\leq Cxp_{i0} \sum_{j=0}^{\lfloor -\frac{\log(xp_{i0})}{\log p}\rfloor} p^{\alpha j} + Cxp_{i1} \sum_{j=0}^{\lfloor -\frac{\log(xp_{i1})}{\log p}\rfloor} p^{\alpha j} + \widetilde{C}x^{1-\alpha} \\
&\leq Cx \sum_{j=0}^{\lfloor -\frac{\log x}{\log p}\rfloor - 1} p^{\alpha j} + \widetilde{C}x^{1-\alpha} \\
&\leq Cx \sum_{j=0}^{\lfloor -\frac{\log x}{\log p}\rfloor} p^{\alpha j}
\end{aligned}$$

where the last inequality holds because $\widetilde{C} \leq C$ and $x^{1-\alpha} \leq xp^{\alpha\lfloor -\log x/\log p\rfloor}$.

Finally, (3.17) implies the assertion since $\sum_{j=0}^\infty p^{\alpha j} = \frac{1}{1-p^\alpha} < \infty$. $\square$

After getting asymptotic results for the *Poissonized* case via lemma 3.2.2, one needs to transfer these results to the original problem analyzing $\mathrm{Var}(X_n^i)$. This transfer is usually called *Depoissonization*.

In fact, bounds on $\mathrm{Var}(X_{N_\lambda}^i)$ imply upper bounds on $\mathbb{E}[V_i(N_\lambda)]$, $V_i(n) := \mathrm{Var}(X_n^i)$, which may be transferred to $\mathrm{Var}(X_n^i)$ by the following lemma:

Lemma 3.2.3. *Let $(a(n))_{n\in\mathbb{N}_0}$ be a real valued sequence. Furthermore, let N_λ be Poisson $\Pi(\lambda)$ distributed for $\lambda > 0$. Then, as $n \to \infty$, $\Delta a(n) = \mathrm{O}(\sqrt{n})$ implies*

$$|a(n) - \mathbb{E}[a(N_n)]| = \mathrm{O}(n).$$

Proof. First note that $\Delta a(n) = \mathrm{O}(\sqrt{n})$ implies that there exists a constant $C > 0$ such that for all $n, m \in \mathbb{N}_0$

$$|a(n) - a(m)| = \left|\sum_{i=m\wedge n}^{m\vee n - 1} \Delta a(i)\right| \leq C\sqrt{n+m}|n-m|.$$

This yields

$$|a(n) - \mathbb{E}[a(N_n)]| \leq \mathbb{E}[|a(n) - a(N_n)|] \leq C\mathbb{E}[\sqrt{n+N_n}|N_n - n|]$$

and the Cauchy-Schwarz inequality implies the assertion. $\square$

3.3 The Contraction Method

The *Contraction Method* was introduced in 1991 by Uwe Rösler [59] for the analysis of the complexity of the *Quicksort* algorithm. There have been several publications [12, 14, 39, 54, 55, 56, 57, 60, 61] in the last two decades that extend the *Contraction Method* to cover many other recursive structures and algorithms. An extension to systems of distributional equation as they appear in the analysis under the *Markov Source Model* was simultaneously developed in [44, 48].

The introduction of the contraction method requires some notation. Throughout this section, let $\mathfrak{P}$ denote the set of all probability distributions on $\mathbb{R}$ and

$$\mathfrak{P}_s := \{\mathcal{L}(Z) \in \mathfrak{P} \; : \; \mathbb{E}[|Z|^s] < \infty\}, \qquad s > 0,$$

the set of all probability distributions on $\mathbb{R}$ with finite s-th moment. For $M_1 \in \mathbb{R}$ and $M_2 \geq 0$ let

$$\mathfrak{P}_s(M_1) := \{\mathcal{L}(Z) \in \mathfrak{P}_s \; : \; \mathbb{E}[Z] = M_1\}, \qquad s \geq 1,$$
$$\mathfrak{P}_s(M_1, M_2) := \{\mathcal{L}(Z) \in \mathfrak{P}_s \; : \; \mathbb{E}[Z] = M_1, \mathbb{E}[Z^2] = M_2\}, \qquad s \geq 2.$$

Finally, let $\|\cdot\|_p$ denote the p-norm on $\mathfrak{P}_p$ which, for $\mathcal{L}(W) \in \mathfrak{P}_p$ and $p > 0$, is given by

$$\|W\|_p := \|\mathcal{L}(W)\|_p := \left(\mathbb{E}\left[|W|^p\right]\right)^{\min\left\{\frac{1}{p},1\right\}}.$$

3.3.1 Contraction Method in the Bernoulli Source Model

The *Contraction Method* is an approach to derive a (weak) limit of a sequence $(Y_n)_{n \geq 0}$ that satisfies some sort of distributional recursion. Under a quite general framework, such a recursion might be

$$Y_n \stackrel{d}{=} \sum_{i=1}^{K} A_n^{(i)} Y_{I_n^{(i)}}^{(i)} + b_n$$

where $(Y_n^{(1)})_{n \geq 0}, \ldots, (Y_n^{(K)})_{n \geq 0}, (A_n^{(1)}, \ldots, A_n^{(K)}, b_n, I_n^{(1)}, \ldots, I_n^{(K)})$ are independent, $I_n^{(i)}$ is a random variable on $\{0, \ldots, n\}$ and $\mathcal{L}(Y_\ell^{(i)}) = \mathcal{L}(Y_\ell)$ for $i \in \{1, \ldots, K\}$ and $\ell \geq 0$. In most applications, K is a fixed integer. However, there are generalizations to integers $K = K_n$ that depend on n and might be random. Also note that Y_n does not need to be real-valued. A general approach (also covering d-dimensional vectors) is given in [54] including several applications in the field of random structures and algorithms. A functional approach (where Y_n might be a continuous or càdlàg function) is given in [56].

The analysis of *Radix Sort* and *Digital Trees* in the *Bernoulli Source Model* is such an application. In this model (embedded in the *Markov Source Model* by choosing $p_{01} = p_{11} =: p$), the system (3.1) is reduced to a single distributional equation for a sequence $(X_n)_{n \geq 0}$ that is

$$X_{n+d} \stackrel{d}{=} X_{I_n} + \widetilde{X}_{n-I_n} + \eta_n, \qquad n \in \mathbb{N}, \qquad (3.19)$$

where $(X_\ell)_{\ell \geq 0}$, $(\widetilde{X}_\ell)_{\ell \geq 0}$ and (I_n, η_n) are independent, $\mathcal{L}(I_n) = B(n, p)$ and $\mathcal{L}(\widetilde{X}_\ell) = \mathcal{L}(X_\ell)$ for all $\ell \geq 0$.

The upcoming step by step introduction sketches the *Contraction Method* approach. There, mean and variance of X_n are abbreviated by

$$\nu(n) := \mathbb{E}[X_n], \quad \sigma(n) := \sqrt{\text{Var}(X_n)}, \quad n \in \mathbb{N}_0.$$

1. Rescaling: Typically, the quantity X_n needs to be rescaled before weak convergence occurs. If a normal limit is expected, exact normalization is necessary. In this case, the rescaled random variable is defined as

$$Y_n := \begin{cases} \frac{X_n - \nu(n)}{\sigma(n)}, & \text{if } \sigma(n) > 0, \\ 0, & \text{otherwise.} \end{cases}$$

In other applications, such as the original work on *Quicksort* [59], a proper rescaling might be "guessed" and Y_n does not have to be centered or normalized. In fact, it depends on the metric chosen in step 4 whether Y_n needs to be centered and normalized or not.

After rescaling, the stochastic recursion (3.19) provides a similar recursion for $(Y_n)_{n \geq 0}$:

$$Y_{n+d} \stackrel{d}{=} A_{n,1} Y_{I_n} + A_{n,2} \widetilde{Y}_{n-I_n} + b_n \qquad (3.20)$$

where $(Y_n)_{n \geq 0}$, $(\widetilde{Y}_n)_{n \geq 0}$ and $(A_{n,1}, A_{n,2}, b_n, I_n)$ are independent, $\mathcal{L}(\widetilde{Y}_n) = \mathcal{L}(Y_n)$ and $\mathcal{L}(I_n) = B(n, p)$. The coefficients are given by

$$A_{n,1} = \frac{\sigma(I_n)}{\sigma(n+d)}, \quad A_{n,2} = \frac{\sigma(n - I_n)}{\sigma(n+d)}, \quad b_n = \frac{\nu(I_n) + \nu(n - I_n) + \eta_n - \nu(n+d)}{\sigma(n+d)}. \qquad (3.21)$$

2. Asymptotic behavior of the coefficients: The asymptotic behavior of the coefficients in (3.20) determines a possible weak limit of Y_n (step 3). Therefore, limits A_1, A_2 and b are required such that

$$(A_{n,1}, A_{n,2}, b_n) \longrightarrow (A_1, A_2, b).$$

The type of convergence that is needed depends on the chosen metric (discussed in step 4). Typically, convergence in $\|\cdot\|_p$ is sufficient for some $p > 2$ and a suitable realization of $(A_{n,1}, A_{n,2}, b_n)_{n \geq 0}$, A_1, A_2, b in a common probability space.

In many applications, these limits are easy to obtain, in particular, if no exact normalization is needed. However, the asymptotic analysis of (3.21) requires the first order asymptotic of the variance and a sufficiently detailed asymptotic expansion of the mean in order to determine the limit of b_n.

A discussion on such an expansion is given in section 3.3.3. With a bounded increments argument on the error term in the asymptotic expansion of $\nu(n)$ one can show that

$$(A_{n,1}, A_{n,2}, b_n) \xrightarrow{L_3} (\sqrt{p}, \sqrt{1-p}, 0).$$

3. The limit equation: The convergence of the coefficients and (3.20) suggest that if Y_n converges to a limit Y, this limit should satisfy

$$Y \stackrel{d}{=} \sqrt{p} Y + \sqrt{1-p} \widetilde{Y}, \qquad (3.22)$$

3.3. THE CONTRACTION METHOD

where Y and $\widetilde{Y}$ are independent and identically distributed. Hence, the limiting distribution $\mathcal{L}(Y)$ should be a fixed point of the following map:

$$T : \mathfrak{P} \longrightarrow \mathfrak{P},$$
$$\rho \mapsto \mathcal{L}\left(\sqrt{p}Z + \sqrt{1-p}\widetilde{Z}\right), \tag{3.23}$$

where Z and $\widetilde{Z}$ are independent and identically distributed with $\mathcal{L}(Z) = \rho$.

4. Choice of a proper metric. The next step is to endow (a subspace of) $\mathfrak{P}$ with a metric d such that T is a contraction with respect to that metric. Since T is an approximation of the stochastic recursion (3.19) for large n, Banach's fixed-point theorem suggests that $(\mathcal{L}(Y_n))_{n \geq 0}$ should converge with respect to d. Convergence in the metric d should at least imply weak convergence. There are two types of metrics used in context of the contraction method: the *Wasserstein* metric ℓ_p, $p > 0$, and the *Zolotarev* metric ζ_s, $s > 0$.

The Wasserstein metric: In his original work [59] on the *Contraction Method*, Uwe Rösler derived a limit for the *Quicksort* complexity in the *Wasserstein* metric. For a given $p > 0$, the *Wasserstein* distance of $\mu, \rho \in \mathfrak{P}_p$ is defined as

$$\ell_p(\mu, \rho) = \inf\{\|W - Z\|_p \; : \; \mathcal{L}(W) = \mu, \mathcal{L}(Z) = \rho\}$$

where the infimum is taken over all random vectors (W, Z) on a common probability space with marginals $\mathcal{L}(W) = \mu$ and $\mathcal{L}(Z) = \rho$. It is a well known fact that there are optimal L_p-couplings, i.e. for every $p > 0$ exists a vector (W_0, Z_0) with marginals $\mathcal{L}(W_0) = \mu$, $\mathcal{L}(Z_0) = \rho$ and

$$\ell_p(\mu, \rho) = \|W_0 - Z_0\|_p.$$

For real valued random variables and $p \geq 1$ the infimum is attained by choosing $(W_0, Z_0) = (F_W^{-1}(U), F_Z^{-1}(U))$ where U is uniformly distributed on $[0, 1]$, F_W and F_Z are the distribution functions of $\mathcal{L}(W)$ and $\mathcal{L}(Z)$ and $F^{-1}(x) := \inf\{y \in \mathbb{R} \; : \; F(y) > x\}$ for $x \in [0, 1]$ and a function $F : \mathbb{R} \to [0, 1]$.

In particular, convergence in ℓ_p implies weak convergence and convergence of the p-th moment. For maps

$$T' : \mathfrak{P}_p \longrightarrow \mathfrak{P}_p,$$
$$\rho \mapsto \mathcal{L}(A_1 Z + A_2 \widetilde{Z} + b)$$

with independent $Z, \widetilde{Z}, (A_1, A_2, b)$ and $\mathcal{L}(Z) = \mathcal{L}(\widetilde{Z}) = \rho$ there is the following bound to verify contraction:

For $\mu, \rho \in \mathfrak{P}_p$ let (W_0, Z_0) and $(\widetilde{W}_0, \widetilde{Z}_0)$ be two independent optimal L_p couplings of μ and ρ and let (A_1, A_2, b) be independent of $(W_0, Z_0, \widetilde{W}_0, \widetilde{Z}_0)$ and distributed as in T'. Then,

$$\ell_p(\mu, \rho) \leq \|A_1 W_0 + A_2 \widetilde{W}_0 + b - (A_1 Z_0 + A_2 \widetilde{Z}_0 + b)\|_p \leq \|A_1\|_p \ell_p(\mu, \rho) + \|A_2\|_p \ell_p(\mu, \rho)$$

which implies that T' is a contraction with respect to ℓ_p if $\|A_1\|_p + \|A_2\|_p < 1$.

For $p = 2$ and $\mu, \nu \in \mathfrak{P}_2(0)$ one also has the better bound

$$(\ell_2(\mu, \rho))^2 \leq \mathbb{E}[(A_1 W_0 + A_2 \widetilde{W}_0 + b - (A_1 Z_0 + A_2 \widetilde{Z}_0 + b))^2]$$
$$= \mathbb{E}[A_1^2](\ell_2(\mu, \rho))^2 + \mathbb{E}[A_2^2](\ell_2(\mu, \rho))^2 + 2\mathbb{E}[A_1 A_2]\underbrace{\mathbb{E}[W_0 - Z_0]}_{=0}\mathbb{E}[\widetilde{W}_0 - \widetilde{Z}_0]$$

which implies that $T'|_{\mathfrak{P}_2(0)}$ is a contraction if $\mathbb{E}[A_1^2] + \mathbb{E}[A_2^2] < 1$.

However, the map T in (3.23) violates both conditions. In fact, T cannot be a contraction on $\mathfrak{P}_p(0)$ with respect to any metric because all distributions $\mathcal{N}(0, \sigma^2) \in \mathfrak{P}_p(0)$, $\sigma \geq 0$, are fixed points of T which is a contradiction to the uniqueness of fixed points in contractions.

Therefore, a different kind of a metric and further restrictions on $\mathfrak{P}_p$ are needed in order to restrict T to a contracting map. The authors in [57] proposed to work with a metric developed by Zolotarev in [70].

The Zolotarev metric: The *Zolotarev* metric ζ_s is an ideal metric of order s. Here, $s \in \mathbb{R}^+$ is a parameter which is typically chosen to be $s \in (0, 3]$ when applying the *Contraction Method*.

Recall that, for $M_1 \in \mathbb{R}$ and $M_2 \geq 0$,

$$\mathfrak{P}_s(M_1) := \{\mathcal{L}(Z) \in \mathfrak{P}_s \; : \; \mathbb{E}[Z] = M_1\}, \qquad s \geq 1,$$

$$\mathfrak{P}_s(M_1, M_2) := \{\mathcal{L}(Z) \in \mathfrak{P}_s \; : \; \mathbb{E}[Z] = M_1, \mathbb{E}[Z^2] = M_2\}, \qquad s \geq 2.$$

For $s = m + \alpha$, $m \in \mathbb{N}_0$ and $\alpha \in (0, 1]$, the *Zolotarev* distance of distributions $\mathcal{L}(W), \mathcal{L}(Z) \in \mathfrak{P}_s$ is defined as

$$\zeta_s(\mathcal{L}(W), \mathcal{L}(Z)) := \sup_{f \in \mathcal{F}_s} |\mathbb{E}[f(W)] - \mathbb{E}[f(Z)]| \qquad (3.24)$$

where

$$\mathcal{F}_s := \{f \in \mathcal{C}^m(\mathbb{R}, \mathbb{R}) : |f^{(m)}(x) - f^{(m)}(y)| \leq |x-y|^\alpha\}$$

is the set of all m-times differential functions whose m-th derivative is Hölder continuous with Hölder exponent α and Hölder constant 1.

A priori, the distance (3.24) might not be well-defined or finite. A Taylor expansion of f reveals that

- for $s \in (0, 1]$, $\zeta_s(\nu, \rho)$ is well-defined by (3.24) and finite if $\nu, \rho \in \mathfrak{P}_s$,
- for $s \in (1, 2]$, $\zeta_s(\nu, \rho)$ is well-defined by (3.24) and finite if $\nu, \rho \in \mathfrak{P}_s(M_1)$ for some $M_1 \in \mathbb{R}$,
- for $s \in (2, 3]$, $\zeta_s(\nu, \rho)$ is well-defined by (3.24) and finite if $\nu, \rho \in \mathfrak{P}_s(M_1, M_2)$ for some $M_1 \in \mathbb{R}$ and $M_2 \geq 0$.

More generally, $\zeta_s(\nu, \rho)$ is well-defined and finite if ν and ρ have finite s-th moment and if the k-th moments of ν and ρ are equal for all $k \in \{1, \ldots, \lceil s \rceil - 1\}$.

The *Zolotarev* distance between random variable W and Z with $\mathbb{E}[|W|^s] < \infty$, $\mathbb{E}[|Z|^s] < \infty$ and $\mathbb{E}[W^k] = \mathbb{E}[Z^k]$ for $k = 1, \ldots, m$ is defined as

$$\zeta_s(W, Z) := \zeta_s(\mathcal{L}(W), \mathcal{L}(Z)).$$

There are several useful properties of the *Zolotarev* metric when deriving limits with the *Contraction Method*. First of all, ζ_s is a metric:

Lemma 3.3.1. *Let $M_1 \in \mathbb{R}$, $M_2 \in \mathcal{R}^+$ and $s \in (0, 3]$. Moreover, let*

$$\mathcal{S} := \begin{cases} \mathfrak{P}_s, & \text{if } s \leq 1, \\ \mathfrak{P}_s(M_1), & \text{if } s \in (1, 2], \\ \mathfrak{P}_s(M_1, M_2), & \text{if } s \in (2, 3]. \end{cases}$$

Then, ζ_s is a metric on $\mathcal{S}$.

3.3. THE CONTRACTION METHOD

Proof. Definition (3.24) immediately yields for any $\mu, \nu, \rho \in \mathcal{S}$

$$\zeta_s(\mu, \mu) = 0, \quad \zeta_s(\mu, \nu) = \zeta_s(\nu, \mu), \quad \zeta_s(\mu, \rho) \leq \zeta_s(\mu, \nu) + \zeta_s(\nu, \rho).$$

It remains to show that

$$\zeta_s(\mu, \nu) = 0 \implies \mu = \nu$$

which can be done by considering the characteristic functions of μ and ν. To this end, let X and Y have distribution μ and ν. For any $x \neq 0$, consider the functions

$$f_x : \mathbb{R} \longrightarrow [-1, 1], \, y \mapsto \sin(xy), \qquad g_x : \mathbb{R} \longrightarrow [-1, 1], \, y \mapsto \cos(xy).$$

The mean value theorem reveals that, for suitable constants $c_x, d_x > 0$ depending on s, $c_x f_x \in \mathcal{F}_s$ and $d_x g_x \in \mathcal{F}_s$ and therefore, by $\zeta_s(\mu, \nu) = 0$,

$$\mathbb{E}[\exp(ixX)] = \frac{1}{d_x}\mathbb{E}[d_x g_x(X)] + \frac{i}{c_x}\mathbb{E}[c_x f_x(X)] = \frac{1}{d_x}\mathbb{E}[d_x g_x(Y)] + \frac{i}{c_x}\mathbb{E}[c_x f_x(Y)] = \mathbb{E}[\exp(ixY)].$$

Hence, μ and ν have the same characteristic functions and thus have to be equal. □

With similar arguments concerning the characteristic functions one obtains that convergence in ζ_s implies weak convergence:

Lemma 3.3.2. *With the notation from lemma 3.3.1, the following holds for any $\rho_n, \rho \in \mathcal{S}$, $n \geq 1$:*

$$\lim_{n \to \infty} \zeta_s(\rho_n, \rho) = 0 \implies \rho_n \xrightarrow{w} \rho.$$

Proof. Convergence in ζ_s implies the (pointwise) convergence of the characteristic functions of ρ_n and ρ by considering the functions presented in the proof of lemma 3.3.1. Thus, Lévy's continuity theorem yields the assertion. □

The next property is crucial when showing that a map T is a contraction with respect to ζ_s. The proof is done in [70, Lemma 3].

Lemma 3.3.3. ζ_s *is ideal of order s, i.e.*

(i) $\zeta_s(cX, cY) = |c|^s \zeta_s(X, Y)$,

(ii) $\zeta_s(X + Z, Y + Z) \leq \zeta_s(X, Y)$

for all $c \in \mathbb{R} \setminus \{0\}$, $\mathcal{L}(X), \mathcal{L}(Y) \in \mathfrak{P}_s$ with $\mathbb{E}[X^k] = \mathbb{E}[Y^k]$ for $k = 1, \ldots, \lceil s \rceil - 1$ and Z independent of (X, Y) with $\mathcal{L}(Z) \in \mathfrak{P}_s$.

An immediate consequence of lemma 3.3.3 is the following result on sums of independent random variables:

Corollary 3.3.4. *Let (X_1, Y_1) and (X_2, Y_2) be two independent random variables in $\mathbb{R}^2$ such that the marginals satisfy, for some $s > 0$,*

$$\mathbb{E}[|X_i|^s] < \infty, \quad \mathbb{E}[|Y_i|^s] < \infty, \quad \mathbb{E}[X_i^k] = \mathbb{E}[Y_i^k], \quad \text{for } i = 1, 2 \text{ and } k \in \{1, \ldots, \lceil s \rceil - 1\}.$$

Then, the Zolotarev distance of the sums satisfies

$$\zeta_s(X_1 + X_2, Y_1 + Y_2) \leq \zeta_s(X_1, Y_1) + \zeta_s(X_2, Y_2).$$

Proof. The triangle inequality and lemma 3.3.3(ii) imply

$$\zeta_s(X_1+X_2, Y_1+Y_2) \leq \zeta_s(X_1+X_2, Y_1+X_2) + \zeta_s(Y_1+X_2, Y_1+Y_2)$$
$$\leq \zeta_s(X_1, Y_1) + \zeta_s(X_2, Y_2)$$

which is the assertion. Note that all ζ_s-distances are well-defined and finite because $X_1 + X_2$, $Y_1 + X_2$, $Y_1 + X_2$, $Y_1 + Y_2$ have finite s-th moments and

$$\mathbb{E}[(X_1+X_2)^k] = \mathbb{E}[(Y_1+X_2)^k] = \mathbb{E}[(Y_1+Y_2)^k] \quad \text{for } k=1,\ldots,m$$

by independence and the assumption on the moments of X_1, X_2, Y_1, Y_2. □

In the applications, it is more convenient to bound some of the *Zolotarev* distances with the *Wasserstein* metric instead. There is the following connection between ζ_s and ℓ_s:

Lemma 3.3.5. *Let X and Y be two real valued random variables that satisfy, for some $s > 1$,*

$$\mathbb{E}[|X|^s] < \infty, \quad \mathbb{E}[|Y|^s] < \infty, \quad \mathbb{E}[X^k] = \mathbb{E}[Y^k], \quad \text{for } k \in \{1, \ldots, \lceil s \rceil - 1\}.$$

Then, the Zolotarev distance of X and Y is bounded by

$$\zeta_s(X,Y) \leq (\mathbb{E}[|X|^s]^{1-1/s} + \mathbb{E}[|Y|^s]^{1-1/s})\ell_s(X,Y) \tag{3.25}$$

where ℓ_s denotes the Wasserstein metric given by

$$\ell_s(X,Y) := \ell_s(\mathcal{L}(X), \mathcal{L}(Y)) := \inf\{\|X'-Y'\|_s \; : \; X' \stackrel{d}{=} X, Y' \stackrel{d}{=} Y\}.$$

Proof. Fix $s > 1$ and an arbitrary $f \in \mathcal{F}_s$. Let $g : \mathbb{R} \to \mathbb{R}$ be defined by

$$g(x) = f(x) - \sum_{k=0}^{m} \frac{f^{(k)}(0)}{k!} x^k$$

where $f^{(k)}$ denotes the k-th derivative of f. Note that

$$g^{(j)}(x) = f^{(j)}(x) - f^{(j)}(0) - \sum_{k=j+1}^{m} \frac{f^{(k)}(0)}{(k-j)!} x^{k-j}, \quad 0 \leq j \leq m$$

which yields $g^{(j)}(0) = 0$ for $j = 0, \ldots, m$ and

$$|g^{(m)}(x) - g^{(m)}(y)| = |f^{(m)}(x) - f^{(m)}(y)| \leq |x-y|^\alpha.$$

Since the mean value theorem implies for any $j \leq m-1$ and $x \in \mathbb{R}$

$$|g^{(j)}(x)| = |g^{(j)}(x) - g^{(j)}(0)| \leq \sup_{|y| \leq x} |g^{(j+1)}(y)| \cdot |x|,$$

one obtains by a backward induction on j that

$$|g^{(j)}(x)| \leq |x|^{s-j}, \quad 0 \leq j \leq m.$$

Hence, $Z := Y - X$ and the mean value theorem yield for some suitable constant $0 \leq \vartheta \leq 1$

$$|g(Y) - g(X)| = |g(X+Z) - g(X)| = |g^{(1)}(X+\vartheta Z)| \cdot |Z|$$
$$\leq |X+\vartheta Z|^{s-1} \cdot |Z| \leq (|X|^{s-1} + |Y|^{s-1})|Z|$$

3.3. THE CONTRACTION METHOD

where the last inequality holds because $|X + \vartheta Z| \leq \max\{|X|, |Y|\}$ and therefore, $|X + \vartheta Z|^{s-1} \leq \max\{|X|^{s-1}, |Y|^{s-1}\}$. Since the first m moments of X and Y are equal, Hölder's inequality implies

$$\begin{aligned}|\mathbb{E}[f(Y) - f(X)]| &= |\mathbb{E}[g(Y) - g(X)]| \\ &\leq \mathbb{E}\left[(|X|^{s-1} + |Y|^{s-1}) \cdot |Z|\right] \\ &= \mathbb{E}[|X|^{s-1}|Z|] + \mathbb{E}[|Y|^{s-1}|Z|] \\ &\leq \mathbb{E}[|X|^s]^{1-1/s}\mathbb{E}[|Z|^s]^{1/s} + \mathbb{E}[|Y|^s]^{1-1/s}\mathbb{E}[|Z|^s]^{1/s}.\end{aligned}$$

Note that this holds for any coupling (X, Y). In particular, if (X, Y) is an optimal L_s-coupling,

$$|\mathbb{E}[f(Y) - f(X)]| \leq \left(\mathbb{E}[|X|^s]^{1-1/s} + \mathbb{E}[|Y|^s]^{1-1/s}\right) \ell_s(X, Y)$$

which yields the assertion by taking the supremum over all $f \in \mathcal{F}_s$. □

5. Limit laws for a contracting map: The map T defined in (3.23) is a contraction on $\mathfrak{P}_s(0,1)$ with respect to the Zolotarev metric ζ_s for $s \in (2,3]$. The contraction property follows by corollary 3.3.4 and lemma 3.3.3(i) which yields for any $\mathcal{L}(W), \mathcal{L}(Z) \in \mathfrak{P}_s(0,1)$

$$\begin{aligned}\zeta_s\big(T(\mathcal{L}(W)), T(\mathcal{L}(Z))\big) &= \zeta_s\big(\sqrt{p}W + \sqrt{1-p}\widetilde{W}, \sqrt{p}Z + \sqrt{1-p}\widetilde{Z}\big) \\ &\leq (\sqrt{p})^s \zeta_s(W, Z) + (\sqrt{1-p})^s \zeta_s(\widetilde{W}, \widetilde{Z}) \\ &= \underbrace{\left(p^{s/2} + (1-p)^{s/2}\right)}_{<1 \text{ for } s>2} \zeta_s(\mathcal{L}(W), \mathcal{L}(Z)).\end{aligned}$$

In particular, $\mathcal{N}(0,1) \subset \mathfrak{P}_s(0,1)$ is the unique fixed point of $T\big|_{\mathfrak{P}_s(0,1)}$. In other applications with non-normal limits, the existence of a fixed point needs to be proven first. For distributions on $\mathbb{R}$, the completeness of $(\mathfrak{P}_s(0,1), \zeta_s)$ implies the existence of the fixed point.

Intuitively, T approximates the distributional recursion (3.20) for large n. Iterating (3.20) on the right hand side therefore suggests, by Banach's fixed point theorem, that $\mathcal{L}(Y_n)$ converges to the unique fixed point $\mathcal{N}(0,1)$ of $T\big|_{\mathfrak{P}_s(0,1)}$. Note that the exact normalization in the definition of Y_n is required to ensure that $\mathcal{L}(Y_n) \in \mathfrak{P}_s(0,1)$ (for all n with $\sigma(n) > 0$).

A rigorous proof in the *Markov Source Model* is done in section 4.3.

3.3.2 Generalization to the Markov Source Model

In the framework presented so far, only a single sequence $(X_n)_{n\geq 0}$ with a distributional recursion is discussed. Recall that in the *Markov Source Model*, sequences $(X_n^0)_{n\geq 0}$ and $(X_n^1)_{n\geq 0}$ appear that satisfy a system of distributional recursions. This system is given by

$$\begin{aligned}X_{n+d}^0 &\stackrel{d}{=} X_{I_n^0}^0 + X_{n-I_n^0}^1 + \eta_n^0, \\ X_{n+d}^1 &\stackrel{d}{=} X_{I_n^1}^0 + X_{n-I_n^1}^1 + \eta_n^1,\end{aligned} \qquad (3.26)$$

with $(X_n^0)_{n\geq 0}$, $(X_n^1)_{n\geq 0}$ and $(I_n^0, I_n^1, \eta_n^0, \eta_n^1)_{n\in\mathbb{N}_0}$ independent and $\mathcal{L}(I_n^i) = B(n, p_{i0})$, $i \in \Sigma$.

The previous strategy (1.-5.) remains the same. Mean and variance of X_n^0 and X_n^1 are abbreviated by
$$\nu_i(n) := \mathbb{E}[X_n^i], \quad V_i(n) := \text{Var}(X_n^i), \quad \sigma_i(n) := \sqrt{V_i(n)}, \quad n \in \mathbb{N}_0, \ i \in \Sigma.$$

1. Rescaling: The study of the *Bernoulli Source Model* suggests that an exact normalization is required to enable finite ζ_s distances for $s \in (2,3]$. Therefore, one considers the normalized sequences $(Y_n^0)_{n \geq 0}$ and $(Y_n^1)_{n \geq 0}$ which, for $n \in \mathbb{N}_0$ and $i \in \Sigma$, are defined as

$$Y_n^i := \begin{cases} \frac{X_n^i - \nu_i(n)}{\sigma_i(n)}, & \text{if } \sigma_i(n) > 0, \\ 0, & \text{otherwise.} \end{cases}$$

System (3.26) leads to a similar system of distributional equations for the rescaled quantities: $(Y_n^0)_{n \geq 0}$ and $(Y_n^1)_{n \geq 0}$ satisfy for all $n \in \mathbb{N}$ with $\sigma_0(n+d) > 0$ and $\sigma_1(n+d) > 0$

$$\begin{aligned} Y_{n+d}^0 &\stackrel{d}{=} A_{n,1}^0 Y_{I_n^0}^0 + A_{n,2}^0 Y_{n-I_n^0}^1 + b_n^0, \\ Y_{n+d}^1 &\stackrel{d}{=} A_{n,1}^1 Y_{I_n^1}^0 + A_{n,2}^1 Y_{n-I_n^1}^1 + b_n^1, \end{aligned} \quad (3.27)$$

where $(Y_n^0)_{n \geq 0}$, $(Y_n^1)_{n \geq 0}$ and $(I_n^i, A_{n,1}^i, A_{n,2}^i, b_n^i)$ are independent for both $i \in \Sigma$.

Here, the coefficients are given by

$$A_{n,1}^i = \frac{\sigma_0(I_n^i)}{\sigma_i(n+d)}, \quad A_{n,2}^i = \frac{\sigma_1(n - I_n^i)}{\sigma_i(n+d)}, \quad b_n^i = \frac{\nu_0(I_n^i) + \nu_1(n - I_n^i) + \eta_n^i - \nu_i(n+d)}{\sigma_i(n+d)}.$$

2. Asymptotic behavior of the coefficients: The study of the asymptotic behavior of the coefficients requires the first order asymptotic of the variances and a sufficiently detailed asymptotic study of the means. The results derived in sections 4.1 and 4.2 lead to a limit that is, as $n \to \infty$,

$$(A_{n,1}^i, A_{n,2}^i, b_n^i) \xrightarrow{L_3} (\sqrt{p_{i0}}, \sqrt{p_{i1}}, 0), \quad i \in \Sigma.$$

3. The limit system. The asymptotic behavior of the coefficients and (3.27) suggest that if $(Y_n^0)_{n \geq 0}$ and $(Y_n^1)_{n \geq 0}$ converge to limits Y^0 and Y^1, these limits should satisfy

$$\begin{aligned} Y^0 &\stackrel{d}{=} \sqrt{p_{00}} Y^0 + \sqrt{p_{01}} Y^1, \\ Y^1 &\stackrel{d}{=} \sqrt{p_{10}} Y^0 + \sqrt{p_{01}} Y^1. \end{aligned}$$

The corresponding limit map is given by

$$\begin{aligned} T : \mathfrak{P} \times \mathfrak{P} &\longrightarrow \mathfrak{P} \times \mathfrak{P}, \\ \begin{pmatrix} \rho_1 \\ \rho_2 \end{pmatrix} &\mapsto \begin{pmatrix} \mathcal{L}(\sqrt{p_{00}} Z_1 + \sqrt{p_{01}} Z_2) \\ \mathcal{L}(\sqrt{p_{10}} Z_1 + \sqrt{p_{11}} Z_2) \end{pmatrix} \end{aligned} \quad (3.28)$$

with Z_1, Z_2 independent and $\mathcal{L}(Z_1) = \rho_1$, $\mathcal{L}(Z_2) = \rho_2$.

4. Choice of a proper metric: The study of the *Bernoulli Source Model* suggests to work with *Zolotarev* distances. A possible generalization of ζ_s to $\mathfrak{P} \times \mathfrak{P}$ is given by the maximum of the distances in each component. More precisely, for $s \in (2,3]$ and $\rho_1, \rho_2, \mu_1, \mu_2 \in \mathfrak{P}_s(0,1)$, let the $\zeta_s^\vee$ distance of (ρ_1, ρ_2) and (μ_1, μ_2) be defined as

$$\zeta_s^\vee \left(\begin{pmatrix} \rho_1 \\ \rho_2 \end{pmatrix}, \begin{pmatrix} \mu_1 \\ \mu_2 \end{pmatrix} \right) := \max \{ \zeta_s(\rho_1, \mu_1), \zeta_s(\rho_2, \mu_2) \}. \quad (3.29)$$

3.3. THE CONTRACTION METHOD

This endows $\mathfrak{P}_s(0,1) \times \mathfrak{P}_s(0,1)$ with a metric. Convergence in $\zeta_s^\vee$ implies weak convergence of each component by lemma 3.3.2.

5. Limit laws for a contracting map: The map T defined in (3.28) is a contraction on $\mathfrak{P}_s(0,1) \times \mathfrak{P}_s(0,1)$ with respect to $\zeta_s^\vee$, $s \in (2,3]$: Note that

$$\zeta_s^\vee\left(T\left(\begin{pmatrix}\rho_1\\\rho_2\end{pmatrix}\right), T\left(\begin{pmatrix}\mu_1\\\mu_2\end{pmatrix}\right)\right) = \max\Big\{\zeta_s(\sqrt{p_{00}}Z_1 + \sqrt{p_{01}}Z_2, \sqrt{p_{00}}W_1 + \sqrt{p_{01}}W_2),$$
$$\zeta_s(\sqrt{p_{10}}Z_1 + \sqrt{p_{11}}Z_2, \sqrt{p_{10}}W_1 + \sqrt{p_{11}}W_2)\Big\}$$

with independent Z_1, Z_2, W_1, W_2 and $\mathcal{L}(Z_i) = \rho_i$, $\mathcal{L}(W_i) = \mu_i$ for $i \in \Sigma$. This yields by lemma 3.3.3 and corollary 3.3.4

$$\zeta_s^\vee\left(T\left(\begin{pmatrix}\rho_1\\\rho_2\end{pmatrix}\right), T\left(\begin{pmatrix}\mu_1\\\mu_2\end{pmatrix}\right)\right) \leq \max\Big\{p_{00}^{s/2}\zeta_s(Z_1, W_1) + p_{01}^{s/2}\zeta_s(Z_2, W_2),$$
$$p_{10}^{s/2}\zeta_s(Z_1, W_1) + p_{11}^{s/2}\zeta_s(Z_2, W_2)\Big\}$$
$$\leq \underbrace{\max\{p_{00}^{s/2} + p_{01}^{s/2}, p_{10}^{s/2} + p_{11}^{s/2}\}}_{<1 \text{ for } s>2}\zeta_s^\vee\left(\begin{pmatrix}\rho_1\\\rho_2\end{pmatrix}, \begin{pmatrix}\mu_1\\\mu_2\end{pmatrix}\right)$$

A complete proof of the convergence of $(\mathcal{L}(Y_n^0), \mathcal{L}(Y_n^1))$ in $\zeta_s^\vee$ is given in section 4.3.

Remark on a multivariate approach: The analysis of the *Markov Source Model* was proposed to be done by considering the system (3.27). One could also consider two dimensional vectors

$$Y_n := \begin{pmatrix} Y_n^0 \\ Y_n^1 \end{pmatrix}$$

by introducing a suitable (in-)dependence between Y_n^0 and Y_n^1 in a common probability space. Any reasonable embedding should lead to the limit equation

$$Y \stackrel{d}{=} A_1 Y + A_2 \widetilde{Y}$$

with $Y, \widetilde{Y}$ independent, $\mathcal{L}(\widetilde{Y}) = \mathcal{L}(Y)$ and 2×2 matrices A_1, A_2 given by

$$A_1 = \begin{pmatrix} \sqrt{p_{00}} & 0 \\ 0 & \sqrt{p_{11}} \end{pmatrix}, \quad A_2 = \begin{pmatrix} 0 & \sqrt{p_{01}} \\ \sqrt{p_{10}} & 0 \end{pmatrix}. \tag{3.30}$$

Such equations are covered by a multivariate approach of the *Contraction Method* presented in [54]. There, a sufficient condition for a ζ_3-contraction is given by [54, condition (25)] which is

$$\|A_1\|_{op}^3 + \|A_2\|_{op}^3 < 1.$$

The operator norms of the matrices (3.30) are

$$\|A_1\|_{op} = \sqrt{\max\{p_{00}, p_{11}\}}, \quad \|A_2\|_{op} = \sqrt{\max\{p_{01}, p_{10}\}}.$$

Hence, [54, condition (25)] requires

$$(\max\{p_{00}, p_{11}\})^{3/2} + (\max\{p_{01}, p_{10}\})^{3/2} < 1.$$

Such an additional restriction on the transition matrix P is avoided when working with the system (3.27) and the $\zeta_s^\vee$ metric.

3.3.3 Lipschitz-Continuity in the Asymptotic Analysis

Deriving limit laws with the *Contraction Method* requires an understanding of the asymptotic behavior of the coefficients that appear after rescaling. Recall that the typical coefficients after normalization are given by

$$A^i_{n,1} = \frac{\sigma_0(I^i_n)}{\sigma_i(n+d)}, \quad A^i_{n,2} = \frac{\sigma_1(n-I^i_n)}{\sigma_i(n+d)}, \quad b^i_n = \frac{\nu_0(I^i_n) + \nu_1(n-I^i_n) + \eta^i_n - \nu_i(n+d)}{\sigma_i(n+d)}.$$

The analysis of $A^i_{n,1}$ and $A^i_{n,2}$ only involves the first order asymptotic of $\sigma_0(n)$ and $\sigma_1(n)$ as $n \to \infty$. In fact, the first order asymptotic of the variance derived in section 4.2 is given by

$$\sigma_i(n) \sim \sigma\sqrt{n \log n}, \quad i \in \Sigma,$$

for some constant $\sigma > 0$. Hence, the law of large numbers yields almost surely, as $n \to \infty$,

$$A^i_{n,1} \longrightarrow \sqrt{p_{i0}}, \quad A^i_{n,2} \longrightarrow \sqrt{p_{i1}}, \quad i \in \Sigma.$$

which may also be extended to convergence in L_p by the dominated convergence theorem. A more detailed discussion on the asymptotic behavior is given in lemma 4.3.2.

The analysis of b^i_n is more involved due to the fact that the complete asymptotic expansion of $\nu_i(n)$ up to the order of $o(\sqrt{n \log n})$ is relevant for the asymptotic behavior of b^i_n. However, even in the study of the path length in *Tries* under the (asymmetric) *Bernoulli Source Model* such an expansion seems far out of reach.

Even very powerful analytical approaches lead to an asymptotic expansion of the mean that is

$$\frac{1}{H} n \log n + \chi(\log n) n + o(n)$$

where χ is some bounded, periodic function (that might be constant depending on the source). Details on a general analytical approach for *Bernoulli Sources* can be found in [19].

Therefore, a different structural property of the error terms

$$f_i(n) = \mathbb{E}[X^i_n] - \frac{1}{H} n \log n, \quad i \in \Sigma,$$

is needed. Such a property should at least imply

$$\left\| \frac{f_0(I^i_n) - \mathbb{E}[f_0(I^i_n)]}{\sqrt{n \log n}} + \frac{f_1(n-I^i_n) - \mathbb{E}[f_1(n-I^i_n)]}{\sqrt{n \log n}} \right\|_3 \longrightarrow 0 \qquad (3.31)$$

in order to derive the asymptotic behavior of b^0_n and b^1_n as $n \to \infty$.

Recall that I^i_n follows the binomial distribution $B(n, p_{i0})$. Hence, any property that keeps the fluctuation of $f_0(I^i_n)$ and $f_1(n-I^i_n)$ of the same order as the fluctuation of I^i_n should help to obtain (3.31).

A useful property in this context is Lipschitz-continuity:

Definition 3.3.6. *Let $S \subset \mathbb{R}$. A function $f : S \to \mathbb{R}$ is called Lipschitz-continuous with Lipschitz-constant C if*

$$|f(x) - f(y)| \leq C|x - y|, \qquad x, y \in S.$$

Moreover, f is called Lipschitz-continuous if there exists a constant $C > 0$ such that f is Lipschitz-continuous with Lipschitz-constant C.

3.3. THE CONTRACTION METHOD

There is one simple criterion for Lipschitz-continuity if the domain of the function is $\mathbb{N}_0$:

Lemma 3.3.7. *Let $f : \mathbb{N}_0 \to \mathbb{R}$ be some function. Then, f is Lipschitz-continuous if and only if the sequence $(|f(n+1) - f(n)|)_{n \in \mathbb{N}_0}$ is bounded.*

Proof. Suppose that the sequence $(|f(n+1) - f(n)|)_{n \in \mathbb{N}_0}$ is bounded by a constant $C > 0$. Then, the differences satisfy for all $m, n \in \mathbb{N}_0$, $m > n$,

$$|f(m) - f(n)| = \left| \sum_{k=n}^{m-1} f(k+1) - f(k) \right| \leq \sum_{k=n}^{m-1} |f(k+1) - f(k)| \leq C|m - n|.$$

Therefore, f is Lipschitz-continuous with Lipschitz-constant C.

On the other hand, if f is Lipschitz-continuous with Lipschitz-constant C, $(|f(n+1) - f(n)|)_{n \in \mathbb{N}_0}$ is obviously bounded by C. □

Indeed, f_0 and f_1 have bounded increments. This is shown in section 4.1 and only requires the transfer lemma 3.1.3.

Finally, the asymptotic behavior of b_n^0 and b_n^1 is derived by the following lemma:

Lemma 3.3.8. *Let $S \subset \mathbb{R}$ and $f : S \to \mathbb{R}$ be a Lipschitz-continuous function with Lipschitz-constant C. Moreover, let X be a S-valued random variable that is p-integrable for some $p \geq 1$. Then,*

$$\|f(X) - \mathbb{E}[f(X)]\|_p \leq 2C \|X - \mathbb{E}[X]\|_p.$$

Proof. Let $\widetilde{X}$ be an independent copy of X. Jensen's inequality implies

$$\|f(X) - \mathbb{E}[f(X)]\|_p = \mathbb{E}[|\mathbb{E}[f(X) - f(\widetilde{X})|X]|^p]^{1/p}$$
$$\leq \mathbb{E}[|f(X) - f(\widetilde{X})|^p]^{1/p}.$$

Moreover, the Lipschitz-continuity of f yields

$$|f(X) - f(\widetilde{X})| \leq C|X - \widetilde{X}|$$

and therefore,

$$\|f(X) - \mathbb{E}[f(X)]\|_p \leq C\mathbb{E}[|X - \widetilde{X}|^p]^{1/p}$$
$$= C\|X - \mathbb{E}[X] - (\widetilde{X} - \mathbb{E}[\widetilde{X}])\|_p$$
$$\leq 2C\|X - \mathbb{E}[X]\|_p.$$

□

Chapter 4
Moments and Limit Theorems

The study of parameters in several structures under the *Markov Source Model* leads to a random variable X_n^μ that satisfies a distributional equation (1.7) given by

$$X_{n+d}^\mu \stackrel{d}{=} X_{K_n^\mu}^0 + X_{n-K_n^\mu}^1 + \eta_n^\mu, \quad n \in \mathbb{N}.$$

Hence, the crucial part in the analysis of X_n^μ is a better understanding of the asymptotic behavior of X_n^0 and X_n^1. Such asymptotic results are entirely based on the system (1.8) which is

$$\begin{aligned} X_{n+d}^0 &\stackrel{d}{=} X_{I_n^0}^0 + X_{n-I_n^0}^1 + \eta_n^0, \\ X_{n+d}^1 &\stackrel{d}{=} X_{I_n^1}^0 + X_{n-I_n^1}^1 + \eta_n^1, \end{aligned} \qquad (4.1)$$

with $(X_0^0, \ldots, X_n^0)$, $(X_0^1, \ldots, X_n^1)$ and (I_n^0, I_n^1) independent, $\mathcal{L}(I_n^i) = B(n, p_{i0})$ and $\eta_n^i = g_n^i(I_n^i)$ for some function $g_n^i : \{0, \ldots, n\} \to \mathbb{R}$, $i \in \Sigma$.

The analysis of $\mathbb{E}[X_n^0]$ and $\mathbb{E}[X_n^1]$ in section 4.1 works with some simple transfer lemmas presented in the previous chapter. The analysis of $\mathrm{Var}(X_n^0)$ and $\mathrm{Var}(X_n^1)$ in section 4.2 requires a more careful use of the transfer results including a *Poissonization* argument and a suitable splitting of the random variables X_n^0 and X_n^1. Finally, a limit law of X_n^0 and X_n^1 is derived via the *Contraction Method* in section 4.3. These results on mean, variance and weak convergence are transferred to X_n^μ for arbitrary μ in section 4.4.

Recall that $\Sigma = \{0, 1\}$ denotes the binary alphabet and X_n^0, X_n^1 satisfy the initial conditions $X_n^0 = X_n^1 = 0$, $n \leq \max\{d, 1\}$, and are s-integrable for some $s \in (2, 3]$ and all $n \in \mathbb{N}_0$.

Also recall that η_n^0 and η_n^1 satisfy the conditions

$$\begin{aligned} \mathbb{E}[\eta_n^i] &= n + \mathrm{O}\left(n^{\frac{1}{2}-\varepsilon}\right), & \mathbb{E}[\Delta \eta_n^i] &= 1 + \mathrm{O}\left(n^{-\varepsilon}\right), \\ \mathrm{Var}(\eta_n^i) &= \mathrm{O}\left(n^{1-\varepsilon}\right), & \mathrm{Var}(\Delta \eta_n^i) &= \mathrm{O}(1), \\ \|\eta_n^i - \mathbb{E}[\eta_n^i]\|_3 &= o(\sqrt{n \log n}), & |\eta_n^i| &\leq Cn \end{aligned} \qquad (4.2)$$

where $\Delta \eta_n^i := \eta_{n+1}^i - \eta_n^i = g_{n+1}^i(I_n^i + J_i) - g_n^i(I_n^i)$ with I_n^i and J_i independent and J_i Bernoulli $B(p_{i0})$ distributed.

4.1 Analysis of the Mean

In addition to a first order asymptotic of the expectation of X_n^0 and X_n^1, the crucial result in this section is a bound on the increments of the error term that occurs in the asymptotic expansion. Such a bound is needed to derive the limit laws in section 4.3.

Note that the only condition in (4.2) that is required for the analysis in this section is

$$\mathbb{E}[\Delta \eta_n^i] = 1 + \mathrm{O}(n^{-\varepsilon}).$$

Recall that $\Delta a_n := a_{n+1} - a_n$ for any real valued sequence $(a_n)_{n \geq 0}$. If more than one index is involved, the difference is defined as $\Delta a(m_n) = a(m_n + 1) - a(m_n)$ for sequences $(a(k))_{k \geq 0}$ and $(m_n)_{n \geq 0}$.

Theorem 4.1.1. *Let $(X_n^0)_{n \geq 0}$ and $(X_n^0)_{n \geq 0}$ be sequences of integrable real valued random variables that satisfy the initial conditions $X_n^0 = X_n^1 = 0$ for $n \leq \max\{d, 1\}$ and the distributional recursions (4.1). Suppose there exists a constant $\varepsilon > 0$ such that for all $i \in \Sigma$*

$$\mathbb{E}[\Delta \eta_n^i] = 1 + \mathrm{O}(n^{-\varepsilon}).$$

Then, the expectation of X_n^i, $i \in \Sigma$, satisfies

$$\mathbb{E}[X_n^i] = \frac{1}{H} n \log n + \mathrm{O}(n).$$

Here, H denotes the Source Entropy defined in (1.4).
More precisely, let $f_0 : \mathbb{N}_0 \to \mathbb{R}$ and $f_1 : \mathbb{N}_0 \to \mathbb{R}$ be given by

$$f_i(n) := \mathbb{E}[X_n^i] - \frac{1}{H} n \log n, \quad i \in \Sigma, \ n \in \mathbb{N}_0.$$

Then, f_0 and f_1 are Lipschitz-continuous, i.e. there exists a constant $C > 0$ such that, for every $i \in \Sigma$ and $n, m \in \mathbb{N}_0$,

$$|f_i(n) - f_i(m)| \leq C|n - m|.$$

Remark 4.1.2. *Theorem 4.1.1 also yields bounds in the case $\mathbb{E}[\Delta \eta_n^i] = c + \mathrm{O}(n^{-\varepsilon})$, $c \in \mathbb{R}\setminus\{0\}$ by applying 4.1.1 to X_n^i/c. In these cases, the expectations are given by*

$$\mathbb{E}[X_n^i] = \frac{c}{H} n \log n + \mathrm{O}(n)$$

with a Lipschitz-continuous error term. Moreover, by similar arguments one also obtains bounds in the case $\mathbb{E}[\Delta \eta_n^i] = \mathrm{O}(n^{-\varepsilon})$ which are

$$\mathbb{E}[X_n^i] = \mathrm{O}(n)$$

and, more precisely, $n \mapsto \mathbb{E}[X_n^i]$ is Lipschitz-continuous. Note, however, that in the latter case one usually has $\mathrm{Var}(X_n^i) = \mathrm{O}(n)$ and the Lipschitz-continuity is not sufficient to derive limit laws for X_n^i with the methods presented in section 4.3.

The analysis of the increments of f_i requires a lemma which is a special case of [63, lemma 2]:

Lemma 4.1.3. *Let $(a(m))_{m \in \mathbb{N}_0}$ be a real valued sequence and I_n follow the binomial distribution $B(n, p)$ with $p \in (0, 1)$ and $n \in \mathbb{N}$. Then,*

$$\Delta \mathbb{E}[a(I_n)] = p \mathbb{E}[\Delta a(I_n)], \quad n \in \mathbb{N}.$$

4.1. ANALYSIS OF THE MEAN

Proof. Note that $I_{n+1} \stackrel{d}{=} I_n + B_{n+1}$ where B_{n+1} is independent of I_n and Bernoulli $B(p)$ distributed. Hence, for all $n \in \mathbb{N}$,

$$\Delta \mathbb{E}[a(I_n)] = \mathbb{E}[a(I_n + B_{n+1})] - a(I_n)]$$
$$= p\mathbb{E}[a(I_n+1) - a(I_n)]$$
$$= p\mathbb{E}[\Delta a(I_n)]$$

which is the assertion. $\square$

Proof of theorem 4.1.1. Let $h : [0,1] \to \mathbb{R}$ be the function $h(x) := x \log x$ with the convention that $h(0) = 0$. Recall that the entropy defined in (1.4) is given by

$$H_i = -\sum_{j \in \Sigma} h(p_{ij}), \quad i \in \Sigma, \qquad H = \pi_0 H_0 + \pi_1 H_1$$

where $\pi = \pi_0 \delta_0 + \pi_1 \delta_1$ denotes the stationary distribution of P given by

$$\pi_0 = \frac{p_{10}}{p_{10} + p_{01}}, \qquad \pi_1 = \frac{p_{01}}{p_{10} + p_{01}}.$$

As shown in lemma 3.0.4, the system (4.1) implies a similar system of equations for $\mathbb{E}[X_n^0]$ and $\mathbb{E}[X_n^1]$. Combined with lemma 3.1.2, this system yields $\mathbb{E}[X_n^i] = \frac{1}{H} n \log n + O(n)$ for both $i \in \Sigma$.

For the Lipschitz-continuity of the error terms, first note that the system of equations for $\mathbb{E}[X_n^0]$ and $\mathbb{E}[X_n^1]$ yields a similar system for $f_i(n) = \mathbb{E}[X_n^i] - \frac{1}{H} h(n)$, $i \in \Sigma$. More precisely, f_0 and f_1 satisfy for all $n \in \mathbb{N}_0$ and $i \in \Sigma$

$$f_i(n+d) = \mathbb{E}[f_0(I_n^i) + f_1(n - I_n^i)] + \mathbb{E}[\eta_n^i] - \frac{1}{H}\left(h(n+d) - \mathbb{E}[h(I_n^i)] - \mathbb{E}[h(n-I_n^i)]\right). \quad (4.3)$$

Also note that the Lipschitz-continuity of f_0 and f_1 is equivalent to (see lemma 3.3.7 for details)

$$\sup_{n \in \mathbb{N}} |\Delta f_0(n)| < \infty, \qquad \sup_{n \in \mathbb{N}} |\Delta f_1(n)| < \infty.$$

Taking the increments in (4.3) yields by lemma 4.1.3 and the assumption on $\mathbb{E}[\Delta \eta_n^i]$ that

$$\Delta f_i(n+d) = p_{i0} \mathbb{E}[\Delta f_0(I_n^i)] + p_{i1} \mathbb{E}[\Delta f_1(n - I_n^i)] + \varepsilon_i(n), \quad n \in \mathbb{N}_0, \quad (4.4)$$

where $\varepsilon_i(n)$, $i \in \Sigma$, $n \in \mathbb{N}_0$ satisfies

$$\varepsilon_i(n) = 1 - \frac{1}{H}\left(\Delta h(n+d) - \Delta \mathbb{E}[h(I_n^i)] - \Delta \mathbb{E}[h(n-I_n^i)]\right) + O(n^{-\varepsilon}).$$

The assertion follows from lemma 3.1.4 (remark 3.1.5) and the following asymptotic expansion of $\varepsilon_i(n)$, $i \in \Sigma$, which is shown next:

$$\varepsilon_i(n) = \pi_{1-i} \frac{H_{1-i} - H_i}{H} + O\left(n^{-\min\{\frac{1}{2}, \varepsilon\}}\right), \quad (n \to \infty). \quad (4.5)$$

Such an expansion requires bounds on $\Delta \mathbb{E}[h(I_n^i)]$ and $\Delta \mathbb{E}[h(n - I_n^i)]$. Since I_n^i follows the binomial distribution $B(n, p_{i0})$, lemma 4.1.3 yields

$$\Delta \mathbb{E}[h(I_n^i)] = p_{i0} \mathbb{E}[h(I_n^i + 1) - h(I_n^i)]$$
$$= p_{i0} \mathbb{E}[\log(I_n^i + 1)] + p_{i0} E[I_n^i (\log(I_n^i + 1) - \log I_n^i)]$$
$$= p_{i0}(\log(n+1) + \log p_{i0} + 1) + p_{i0} \mathbb{E}\left[\log\left(\frac{I_n^i + 1}{n+1}\right) - \log p_{i0}\right]$$
$$+ p_{i0} \left(\mathbb{E}[I_n^i(\log(I_n^i + 1) - \log I_n^i)] - 1\right).$$

Note that Chernoff's bound on the tail of the binomial distribution and a Taylor expansion of the logarithm reveal that the remaining expectations are negligible as $n \to \infty$. More precisely, a simple calculation that is given in lemma A.2.2 in the appendix shows that

$$\Delta \mathbb{E}[h(I_n^i)] = p_{i0}(\log(n+1) + \log p_{i0} + 1) + O\left(n^{-\frac{1}{2}}\right).$$

A similar result also holds for the binomial $B(n, p_{i1})$ distributed $n - I_n^i$. Hence,

$$\Delta \mathbb{E}[h(n - I_n^i)] = p_{i1}(\log(n+1) + \log p_{i1} + 1) + O\left(n^{-\frac{1}{2}}\right).$$

Also note that

$$\log(n+d+1) - \log(n+d) = \log\left(1 + \frac{1}{n+d}\right) = \frac{1}{n+d} + O\left(n^{-2}\right)$$

which yields for $\varepsilon_i(n)$, as $n \to \infty$,

$$\varepsilon_i(n) = 1 - \frac{1}{H}(\log(n+d+1) + 1 - h(p_{i0}) - h(p_{i1}) - \log(n+1) - 1) + O\left(n^{-\min\{\frac{1}{2}, \varepsilon\}}\right)$$

$$= 1 - \frac{H_i}{H} + O\left(n^{-\min\{\frac{1}{2}, \varepsilon\}}\right)$$

$$= \pi_{1-i} \frac{H_{1-i} - H_i}{H} + O\left(n^{-\min\{\frac{1}{2}, \varepsilon\}}\right).$$

Finally, note that

$$p_{10}\pi_1 \frac{H_1 - H_0}{H} + p_{01}\pi_0 \frac{H_0 - H_1}{H} = \frac{p_{10}p_{01}}{p_{10} + p_{01}}\left(\frac{H_1 - H_0}{H} + \frac{H_0 - H_1}{H}\right) = 0$$

which in combination with remark 3.1.5 yields the assertion. □

4.2 Analysis of the Variance

The analysis of the variance requires some of the notation from the previous section. Recall that $\pi = \pi_0 \delta_0 + \pi_1 \delta_1$ denotes the stationary distribution of P and that the *Source Entropy* is defined as

$$H = \sum_{i \in \Sigma} \pi_i H_i, \qquad H_i = -\sum_{j \in \Sigma} p_{ij} \log(p_{ij}), \quad i \in \Sigma.$$

The main result in this section is the following theorem:

Theorem 4.2.1. *Let $(X_n^0)_{n \geq 0}$ and $(X_n^1)_{n \geq 0}$ be sequences of real valued random variables with finite mean and variance that satisfy the initial conditions $X_n^0 = X_n^1 = 0$ for $n \leq \max\{d, 1\}$ and the distributional recursions (4.1). Assume that the toll functions η_n^0 and η_n^1 in (4.1) satisfy the conditions (4.2). Then, as $n \to \infty$,*

$$\operatorname{Var}(X_n^i) = \sigma^2 n \log n + O\left(n \sqrt{\log n}\right), \quad i \in \Sigma, \tag{4.6}$$

where the constant σ^2 is given by

$$\sigma^2 = \frac{\pi_0 p_{00} p_{01}}{H^3}\left(\log(p_{00}/p_{01}) + \frac{H_1 - H_0}{p_{01} + p_{10}}\right)^2 + \frac{\pi_1 p_{10} p_{11}}{H^3}\left(\log(p_{10}/p_{11}) + \frac{H_1 - H_0}{p_{01} + p_{10}}\right)^2. \tag{4.7}$$

In particular, $\sigma^2 > 0$ holds for any transition matrix P that satisfies (1.9).

4.2. ANALYSIS OF THE VARIANCE

The proof of theorem 4.2.1 is a bit more complicated than the analysis of the mean. First note that, by lemma 3.0.4, the variances $V_i(n) := \mathrm{Var}(X_n^i)$, $n \in \mathbb{N}_0$, $i \in \Sigma$ satisfy

$$V_i(n+d) = \mathbb{E}[V_0(I_n^i)] + \mathbb{E}[V_1(n - I_n^i)] + \mathrm{Var}(\nu_0(I_n^i) + \nu_1(n - I_n^i) + \eta_n^i)$$

where $\nu_i(n) := \mathbb{E}[X_n^i]$. This suggests a similar approach to the one in section 4.1. However, the expansions derived for ν_0 and ν_1 in theorem 4.1.1 only yield that

$$\mathrm{Var}(\nu_0(I_n^i) + \nu_1(n - I_n^i) + \eta_n^i) = O(n).$$

Hence, it is not directly possible to apply lemma 3.1.2 to derive the first order asymptotic of the variances. A direct application of lemma 3.1.1 only yields that $\mathrm{Var}(X_n^i) = O(n \log n)$, $i \in \Sigma$. In order to apply lemma 3.1.2 it is necessary to derive the first order asymptotic of $\mathrm{Var}(\nu_0(I_n^i) + \nu_1(n - I_n^i) + \eta_n^i)$ which requires an expansion of ν_0 and ν_1 up to an error term of order $O(n^{\frac{1}{2}-\delta})$ for some $\delta > 0$. As already discussed in section 3.3.3, such an asymptotic expansion presently seems to be far out of reach.

The approach presented in this section avoids this problem. Motivated by an approach in [63] which deals with the *Bernoulli Source Model*, X_n^i is split into a sum $Y_n^i + Z_n^i$ in a way that $\mathbb{E}[Y_n^i]$ has a closed form that allows the derivation of the first order asymptotic of $\mathrm{Var}(Y_n^i)$ and that $\mathrm{Var}(Z_n^i) = o(\mathrm{Var}(Y_n^i))$.

To this end, let $(Y_n^i, Z_n^i)_{n \geq 0}$, $i \in \Sigma$ be sequences of pairs of real valued random variables with finite second moments that satisfy the initial conditions

$$Y_n^i = Z_n^i = 0, \qquad n \leq \max\{d,1\}, i \in \Sigma, \tag{4.8}$$

as well as for all $n \geq 1$ and $i \in \Sigma$ the distributional recursions

$$\begin{pmatrix} Y_{n+d}^i \\ Z_{n+d}^i \end{pmatrix} \stackrel{d}{=} \begin{pmatrix} Y_{I_n^i}^0 \\ Z_{I_n^i}^0 \end{pmatrix} + \begin{pmatrix} Y_{n-I_n^i}^1 \\ Z_{n-I_n^i}^1 \end{pmatrix} + \begin{pmatrix} \eta_n^{i,1} \\ \eta_n^{i,2} \end{pmatrix}, \tag{4.9}$$

where $(Y_n^0, Z_n^0)_{n \geq 0}$, $(Y_n^1, Z_n^1)_{n \geq 0}$ and (I_n^0, I_n^1) are independent, $(\eta_n^{i,1})_{n \geq 0}$ is some real valued sequence and $\eta_n^{i,2} = \eta_n^i - \eta_n^{i,1}$, $i \in \Sigma$.

A discussion on the existence of sequences $(Y_n^i, Z_n^i)_{n \geq 0}$, $i \in \Sigma$ with finite second moments that satisfy (4.8) and (4.9) is done in lemma 4.2.9 at the end of the section.

Note that the sums $S_n^i := Y_n^i + Z_n^i$, $i \in \Sigma$, satisfy the initial conditions

$$S_n^i = 0, \qquad n \leq \max\{d,1\}, i \in \Sigma,$$

and moreover, by (4.9) for all $n \geq 1$ the stochastic recursions

$$S_{n+d}^0 \stackrel{d}{=} S_{I_n^0}^0 + S_{n-I_n^0}^1 + \eta_n^0,$$

$$S_{n+d}^1 \stackrel{d}{=} S_{I_n^1}^0 + S_{n-I_n^1}^1 + \eta_n^1.$$

The initial conditions and the system of distributional recursions uniquely define $\mathbb{E}[(S_n^0)^m]$ and $\mathbb{E}[(S_n^1)^m]$, $n \geq 0$, for all $m \in \mathbb{N}$ such that $\mathbb{E}[(S_n^i)^m]$ is finite for all $n \in \mathbb{N}$, $i \in \Sigma$. This can be shown by an induction on m and n which is done for mean and variance in lemma 4.2.9 at the end of the section.

Hence, the variance of X_n^i coincides with the variance of $Y_n^i + Z_n^i$. The proof of theorem 4.2.1 is done by determining the first order asymptotic of Y_n^i, bounding the variance of Z_n^i and combining these results via the next lemma:

Lemma 4.2.2. *Let X and Y be two real valued random variables with a finite second moment. Then,*

$$\left(\sqrt{\text{Var}(X)} - \sqrt{\text{Var}(Y)}\right)^2 \leq \text{Var}(X+Y) \leq \left(\sqrt{\text{Var}(X)} + \sqrt{\text{Var}(Y)}\right)^2. \quad (4.10)$$

In particular, if sequences $(X_n)_{n\geq 0}, (Y_n)_{n\geq 0}$ of real valued random variables with finite second moments satisfy $\text{Var}(Y_n) = o(\text{Var}(X_n))$, then, as $n \to \infty$,

$$\text{Var}(X_n + Y_n) = \text{Var}(X_n) + \text{O}\left(\sqrt{\text{Var}(X_n)\text{Var}(Y_n)}\right). \quad (4.11)$$

Proof. The Cauchy-Schwarz inequality yields

$$|\text{Cov}(X,Y)| \leq \sqrt{\text{Var}(X)}\sqrt{\text{Var}(Y)}$$

which together with $\text{Var}(X+Y) = \text{Var}(X) + \text{Var}(Y) + 2\text{Cov}(X,Y)$ implies (4.10). Moreover, (4.10) obviously implies (4.11). □

Now, the crucial part is to choose $\eta_n^{i,1}$ in a way that for all $n \in \mathbb{N}_0$ and $i \in \Sigma$

(a) $\mathbb{E}[Y_n^i]$ is easy to compute in order to deduce the asymptotics of $\text{Var}(Y_n^i)$,

(b) $\eta_n^{i,1} = n + \text{O}(n^{1/2-\varepsilon})$ which implies $\mathbb{E}[\eta_n^{i,2}] = \text{O}(n^{1/2-\varepsilon})$ and therefore $\text{Var}(Z_n^i) = \text{O}(n)$.

First note that (b), lemma 3.0.4 and transfer lemma 3.1.1 yield $\mathbb{E}[Z_n^i] = \text{O}(n)$ for both $i \in \Sigma$. Since $\mathbb{E}[Y_n^i + Z_n^i] = \mathbb{E}[X_n^i]$, $\eta_n^{i,1}$ has to be chosen in a way that $\mathbb{E}[Y_n^i] = \frac{1}{H}n\log n + \text{O}(n)$. Now let $h(n) := n\log n$ for $n \in \mathbb{N}_0$. A choice of $\eta_n^{i,1}$ that leads to $\mathbb{E}[Y_n^i] = \frac{h(n)}{H}$ is given by

$$\eta_n^{i,1} = \frac{1}{H}(h(n+d) - \mathbb{E}[h(I_n^i) + h(n - I_n^i)])$$

because the function $\frac{h(n)}{H}$ solves the recursions occurring for $\mathbb{E}[Y_n^i]$ (with some minor adjustments in order to include the initial conditions $\mathbb{E}[Y_n^i] = 0$ for $n \leq \max\{d,1\}$).
However, the choice $\frac{1}{H}(h(n+d) - \mathbb{E}[h(I_n^i) + h(n - I_n^i)]) \sim \frac{H_i}{H}n$ still violates condition (b) for $p_{10} \neq p_{00}$ (i.e. except for the *Bernoulli Source Model*).
A choice for $\eta_n^{i,1}$ that leads to $\mathbb{E}[Y_n^i] = \frac{1}{H}n\log n + c_i n$ for some constants c_0 and c_1 is given by

$$\eta_n^{i,1} = \frac{1}{H}(h(n+d) - \mathbb{E}[h(I_n^i) + h(n - I_n^i)]) + c_i(n+d) - c_0 p_{i0} n - c_1 p_{i1} n.$$

Such a toll term satisfies $\eta_n^{i,1} \sim \frac{H_i}{H}n + p_{i\,1-i}(c_i - c_{1-i})n$. Hence, condition (b) holds when $c_i = -\frac{H_i}{(p_{01}+p_{10})H}$ for both $i \in \Sigma$ (recall that $H = \pi_0 H_0 + \pi_1 H_1$ with $\pi_0 = p_{10}/(p_{01}+p_{10})$ and $\pi_1 = 1 - \pi_0$).
The approach given above ignores the (asymptotically negligible) effect of the initial conditions $\mathbb{E}[Y_n^i] = 0$ for $n \leq \max\{d,1\}$. In order to cover this condition and to obtain $\mathbb{E}[Y_n^i] = \frac{h(n)}{H} + c_i n$ for $n \geq \max\{d,1\} + 1$, the toll terms need to be chosen as follows:
For all $n \geq 1$ and $i \in \Sigma$ let

$$\eta_n^{i,1} := \begin{cases} 0, & \text{if } n = 1, d = 0, \\ \frac{h(n+d) - \mathbb{E}[h(I_n^i) + h(n - I_n^i)]}{H} + \pi_{1-i}\frac{H_{1-i} - H_i}{H}n + \delta_i(n), & \text{otherwise} \end{cases} \quad (4.12)$$

4.2. ANALYSIS OF THE VARIANCE

with functions h, δ_0 and δ_1 given by $h(n) := n \log n$ and

$$\delta_i(n) = \frac{\mathbb{E}[h(I_n^i)\mathbb{1}_{\{I_n^i \leq \max\{d,1\}\}} + h(n - I_n^i)\mathbb{1}_{\{n - I_n^i \leq \max\{d,1\}\}}]}{H}$$

$$- \frac{H_0 \mathbb{E}[I_n^i \mathbb{1}_{\{I_n^i \leq \max\{d,1\}\}}] + H_1 \mathbb{E}[(n - I_n^i)\mathbb{1}_{\{n - I_n^i \leq \max\{d,1\}\}}]}{(p_{01} + p_{10})H}$$

$$- \frac{H_i}{(p_{01} + p_{10})H} d.$$

This choice leads to the following result on the mean of Y_n^i which, in particular, paves the way for a result on the first order asymptotic of the variance.

Lemma 4.2.3. *With the choice (4.12), the expectations $\nu_Y^i(n) := \mathbb{E}[Y_n^i]$, $i \in \Sigma$, are given by*

$$\nu_Y^i(n) = \begin{cases} 0, & \text{if } n \leq \max\{d,1\}, \\ \frac{1}{H} n \log n - \frac{H_i}{(p_{01}+p_{10})H} n, & \text{if } n \geq \max\{d,1\} + 1. \end{cases} \quad (4.13)$$

Proof. Recall that $Y_n^0 = Y_n^1 = 0$ for $n \leq \max\{d,1\}$ by the initial conditions (4.8). Moreover, note that the recursions (4.9) imply that $(Y_n^0)_{n \geq 0}$ and $(Y_n^1)_{n \geq 0}$ satisfy for all $n \geq 1$

$$Y_{n+d}^0 \stackrel{d}{=} Y_{I_n^0}^0 + Y_{n-I_n^0}^1 + \eta_n^{0,1},$$

$$Y_{n+d}^1 \stackrel{d}{=} Y_{I_n^1}^0 + Y_{n-I_n^1}^1 + \eta_n^{1,1},$$

with $(Y_n^0)_{n \geq 0}$, $(Y_n^1)_{n \geq 0}$ and (I_n^0, I_n^1) independent. As shown in lemma 3.0.4, this implies that the expectations satisfy for all $n \geq 1$ (recall that $\eta_n^{0,1}$ and $\eta_n^{1,1}$ are not random)

$$\nu_Y^0(n+d) = \mathbb{E}[\nu_Y^0(I_n^0)] + \mathbb{E}[\nu_Y^1(n - I_n^0)] + \eta_n^{0,1},$$

$$\nu_Y^1(n+d) = \mathbb{E}[\nu_Y^0(I_n^1)] + \mathbb{E}[\nu_Y^1(n - I_n^1)] + \eta_n^{1,1}.$$

These recursions and the initial conditions $\nu_Y^0(n) = \nu_Y^1(n) = 0$ for $n \leq \max\{d,1\}$ uniquely define ν_Y^0 and ν_Y^1 which is obvious for $d \geq 1$ and easy to show for $d = 0$. A proof of the uniqueness is given for the sum $Y_n^i + Z_n^i$ in the second part of lemma 4.2.9 at the end of the section. Essentially the same proof also holds for Y_n^i.

Now let $\varphi_i : \mathbb{N}_0 \to \mathbb{R}$, $i \in \Sigma$ be defined as

$$\varphi_i(n) = \begin{cases} 0, & \text{if } n \leq \max\{d,1\}, \\ \frac{1}{H} n \log n - \frac{H_i}{(p_{01}+p_{10})H} n, & \text{if } n \geq \max\{d,1\} + 1. \end{cases}$$

Note that $\eta_n^{0,1}$ and $\eta_n^{1,1}$ are chosen in such a way that for all $n \geq 1$

$$\varphi_0(n+d) = \mathbb{E}[\varphi_0(I_n^0)] + \mathbb{E}[\varphi_1(n - I_n^0)] + \eta_n^{0,1},$$

$$\varphi_1(n+d) = \mathbb{E}[\varphi_0(I_n^1)] + \mathbb{E}[\varphi_1(n - I_n^1)] + \eta_n^{1,1},$$

which together with the initial conditions $\varphi_i(n) = 0 = \nu_Y^i(n)$, $i \in \Sigma$, implies $\nu_Y^i = \varphi_i$. The assertion follows by the definition of φ_0 and φ_1. $\square$

Lemma 4.2.4. *The variance of Y_n^0 and Y_n^1 satisfies, as $n \to \infty$,*

$$\mathrm{Var}(Y_n^i) = \sigma^2 n \log n + O(n), \quad i \in \Sigma,$$

where σ^2 is given by

$$\sigma^2 = \frac{\pi_0 p_{00} p_{01}}{H^3}\left(\log(p_{00}/p_{01}) + \frac{H_1 - H_0}{p_{01} + p_{10}}\right)^2 + \frac{\pi_1 p_{10} p_{11}}{H^3}\left(\log(p_{10}/p_{11}) + \frac{H_1 - H_0}{p_{01} + p_{10}}\right)^2.$$

Proof. Recall that Y_n^0 and Y_n^1 satisfy the initial conditions $Y_n^0 = Y_n^1 = 0$ for $n \leq \max\{d, 1\}$ and, for all $n \geq 1$, the distributional recursions

$$Y_{n+d}^0 \stackrel{d}{=} Y_{I_n^0}^0 + Y_{n-I_n^0}^1 + \eta_n^{0,1},$$

$$Y_{n+d}^1 \stackrel{d}{=} Y_{I_n^1}^0 + Y_{n-I_n^1}^1 + \eta_n^{1,1}.$$

Let mean and variance of Y_n^i be denoted by

$$\nu_Y^i := \mathbb{E}[Y_n^i], \qquad V_Y^i(n) := \text{Var}(Y_n^i), \quad n \in \mathbb{N}_0, i \in \Sigma.$$

Then, lemma 3.0.4 implies for all $n \geq 1$ and $i \in \Sigma$

$$V_Y^i(n+d) = \mathbb{E}[V_Y^0(I_n^i)] + \mathbb{E}[V_Y^1(n - I_n^i)] + \text{Var}(\nu_Y^0(I_n^i) + \nu_Y^1(n - I_n^i) + \eta_n^{i,1}). \tag{4.14}$$

Lemma 3.1.2 shows how to transfer the first order asymptotic of $\text{Var}(\nu_Y^0(I_n^i) + \nu_Y^1(n - I_n^i) + \eta_n^{i,1})$ into an asymptotic result concerning V_Y^i. Hence, it only remains to calculate the first order asymptotic of $\text{Var}(\nu_Y^0(I_n^i) + \nu_Y^1(n - I_n^i) + \eta_n^{i,1})$ for both $i \in \Sigma$.

Note that $\eta_n^{i,1}$ is not random by the choice (4.12), i.e. $\text{Var}(\eta_n^{i,1}) = 0$ for all $n \in \mathbb{N}_0$, $i \in \Sigma$. Therefore, with the notation $h(n) := n \log n$ lemma 4.2.3 yields

$$\text{Var}(\nu_Y^0(I_n^i) + \nu_Y^1(n - I_n^i) + \eta_n^{i,1})$$

$$= \text{Var}\left(\frac{h(I_n^i) + h(n - I_n^i)}{H} - \frac{H_0}{(p_{01} + p_{10})H}I_n^i - \frac{H_1}{(p_{01} + p_{10})H}(n - I_n^i) + R_n^i\right)$$

with an error term R_n^i that occurs because $\nu_Y^i(n) = 0$ for $n \leq \max\{d, 1\}$. This error term is given by

$$R_n^i = -\left(\frac{h(I_n^i)}{H} - \frac{H_0}{(p_{01} + p_{10})H}I_n^i\right) \mathbb{1}_{\{I_n^i \leq \max\{d,1\}\}}$$

$$- \left(\frac{h(n - I_n^i)}{H} - \frac{H_1}{(p_{01} + p_{10})H}(n - I_n^i)\right) \mathbb{1}_{\{n - I_n^i \leq \max\{d,1\}\}}.$$

Note that R_n^i is bounded by $C(\mathbb{1}_{\{I_n^i \leq \max\{d,1\}\}} + \mathbb{1}_{\{n - I_n^i \leq \max\{d,1\}\}})$ for a constant

$$C = \frac{h(\max\{d, 1\})}{H} + \frac{H_0 + H_1}{(p_{01} + p_{10})H} \max\{d, 1\}$$

and therefore, $\text{Var}(R_n^i) = \text{O}(\mathbb{P}(I_n^i \in [0, \max\{d, 1\}] \cup [n - \max\{d, 1\}, n]))$ which together with a standard Chernoff bound on I_n^i given in lemma A.1.1 implies

$$\text{Var}(R_n^i) = \text{O}\left(e^{-p_\wedge^2 n}\right) \tag{4.15}$$

with $p_\wedge = \min\{p_{kl} : k, l \in \Sigma\} > 0$ on assumption (1.9). Next, note that

$$h(I_n^i) + h(n - I_n^i) = I_n^i \log(I_n^i/n) + (n - I_n^i) \log(1 - I_n^i/n) + h(n)$$

$$= I_n^i \log(p_{i0}/p_{i1}) + n \log p_{i1} + I_n^i(\log(I_n^i/n) - \log p_{i0})$$

$$+ (n - I_n^i)(\log(1 - I_n^i/n) - \log p_{i1}) + h(n)$$

4.2. ANALYSIS OF THE VARIANCE

which leads to

$$\operatorname{Var}(\nu_Y^0(I_n^i) + \nu_Y^1(n - I_n^i) + \eta_n^{i,1}) = \operatorname{Var}\left(\frac{1}{H}\left(\log(p_{i0}/p_{i1}) - \frac{H_1 - H_0}{p_{01} + p_{10}}\right)I_n^i + \widetilde{R}_n^i + R_n^i\right)$$

with a second error term $\widetilde{R}_n^i$ that is given by

$$\widetilde{R}_n^i = \frac{I_n^i(\log(I_n^i/n) - \log p_{i0}) + (n - I_n^i)(\log(1 - I_n^i/n) - \log p_{i1})}{H}.$$

An easy calculation reveals (details are given in lemma A.2.1 in the appendix)

$$\operatorname{Var}(I_n^i(\log(I_n^i/n) - \log p_{i0}) + (n - I_n^i)(\log(1 - I_n^i/n) - \log p_{i1})) = \mathrm{O}(\log n)$$

and therefore,

$$\operatorname{Var}(\widetilde{R}_n^i) = \mathrm{O}(\log n).$$

This bound combined with (4.15), $\operatorname{Var}(I_n^i) = np_{i0}p_{i1}$ and lemma 4.2.2 yields

$$\operatorname{Var}(\nu_Y^0(I_n^i) + \nu_Y^1(n - I_n^i) + \eta_n^{i,1}) = \frac{1}{H^2}\left(\log(p_{i0}/p_{i1}) + \frac{H_1 - H_0}{p_{01} + p_{10}}\right)^2 p_{i0}p_{i1}n + \mathrm{O}\left(\sqrt{n \log n}\right)$$

and the assertion follows by (4.14) and lemma 3.1.2. □

The analysis of $\operatorname{Var}(Z_n^i)$ requires a more detailed result on mean and variance of $\eta_n^{i,2}$ for both $i \in \Sigma$. Asymptotic results on $\mathbb{E}[\eta_n^{i,2}]$ and $\operatorname{Var}(\eta_n^{i,2})$ are deduced from the assumption on η_n^i and the following result on $\eta_n^{i,1}$:

Lemma 4.2.5. *The choice (4.12) yields for both $i \in \Sigma$, as $n \to \infty$,*

$$\eta_n^{i,1} = n + \mathrm{O}(n^{1/3}).$$

Proof. Let $h : \mathbb{R}_0^+ \to \mathbb{R}$ be defined as $h(x) = x \log x$ for $x > 0$ and $h(0) = 0$. Recall that

$$H_i := -\sum_{j \in \Sigma} h(p_{ij}), \ i \in \Sigma, \qquad H := \pi_0 H_0 + \pi_1 H_1,$$

where $\pi = \pi_0 \delta_0 + \pi_1 \delta_1$ is the stationary distribution of the Markov chain which is given by

$$\pi_0 = \frac{p_{10}}{p_{10} + p_{01}}, \qquad \pi_1 = p_{01}p_{10} + p_{01}.$$

Note that a standard Chernoff bound on the binomial distribution given in lemma A.1.1 implies for $\delta_0(n)$ and $\delta_1(n)$ in (4.12) that

$$\delta_i(n) = \mathrm{O}(1), \quad i \in \Sigma.$$

Moreover, a Taylor expansion on the logarithm reveals

$$h(n + d) - h(n) = d \log n + n \left(\log(n + d) - \log n\right) = \mathrm{O}(\log n).$$

Hence, it is sufficient to show that

$$\frac{h(n) - \mathbb{E}[h(I_n^i) + h(n - I_n^i)]}{H} + \pi_{1-i}\frac{H_{1-i} - H_i}{H}n = n + \mathrm{O}(n^{1/3}).$$

The identity $h(n) = \mathbb{E}[I_n^i \log n + (n - I_n^i) \log n]$ yields for all $n \in \mathbb{N}$ and $i \in \Sigma$

$$h(n) - \mathbb{E}[h(I_n^i) + h(n - I_n^i)] = -\mathbb{E}[I_n^i \log(I_n^i/n) + (n - I_n^i) \log(1 - I_n^i/n)]$$
$$= nH_i - n\mathbb{E}[h(I_n^i/n) - h(p_{i0}) + h(1 - I_n^i/n) - h(p_{i1})].$$

An easy calculation including the concentration of the binomial distribution and the Taylor expansion of the logarithm reveals that

$$\mathbb{E}[h(I_n^i/n) - h(p_{i0})] = O(n^{-2/3})$$

and similarly, since $n - I_n^i$ follows the binomial distribution $B(n, p_{i1})$,

$$\mathbb{E}[h(1 - I_n^i/n) - g(p_{i1})] = O(n^{-2/3}).$$

Details on the calculation are given in lemma A.2.2 in the appendix.

These bounds imply

$$\frac{h(n) - \mathbb{E}[h(I_n^i) + h(n - I_n^i)]}{H} + \pi_{1-i} \frac{H_{1-i} - H_i}{H} n = \frac{H_i}{H} n + \pi_{1-i} \frac{H_{1-i} - H_i}{H} n + O(n^{1/3})$$
$$= n + O\left(n^{1/3}\right)$$

which yields the assertion. $\square$

Corollary 4.2.6. *Assume that η_m^0 and η_m^1 satisfy the conditions (4.2). Then, the choice (4.12) and $\eta_m^{i,2} = \eta_m^i - \eta_m^{i,1}$ imply for both $i \in \Sigma$, as $n \to \infty$,*

$$\mathbb{E}[\eta_m^{i,2}] = O(n^{1/2-\varepsilon}), \qquad \mathrm{Var}(\eta_m^{i,2}) = O(n^{1-\varepsilon}), \qquad \mathrm{Var}(\Delta \eta_m^{i,2}) = O(1),$$

where $0 < \varepsilon \leq 1/6$ is the constant in (4.2).

Proof. Recall that $\eta_m^{i,1}$ is not random and therefore,

$$\mathbb{E}[\eta_m^{i,1}] = \eta_m^{i,1}, \qquad \mathrm{Var}(\eta_m^{i,1}) = 0.$$

This yields for all $n \in \mathbb{N}$ and $i \in \Sigma$

$$\mathbb{E}[\eta_m^{i,2}] = \mathbb{E}[\eta_m^i] - \eta_m^{i,1}, \qquad \mathrm{Var}(\eta_m^{i,2}) = \mathrm{Var}(\eta_m^i), \qquad \mathrm{Var}(\Delta \eta_m^{i,2}) = \mathrm{Var}(\Delta \eta_m^i)$$

and the assertion follows from the assumption (4.2) on η_m^i and lemma 4.2.5. $\square$

The next step in the analysis of the variance is an upper bound on mean and variance of Z_n^i:

Lemma 4.2.7. *Let $\nu_Z^i : \mathbb{N}_0 \to \mathbb{R}$, $i \in \Sigma$, be defined as*

$$\nu_Z^i(n) := \mathbb{E}[Z_n^i].$$

Then, the functions ν_Z^0 and ν_Z^1 are Lipschitz-continuous, i.e. a constant $C > 0$ exists in a way that, for all $n, m \in \mathbb{N}_0$ and $i \in \Sigma$,

$$|\nu_Z^i(n) - \nu_Z^i(m)| \leq C|n - m|.$$

4.2. ANALYSIS OF THE VARIANCE

Proof. Recall that
$$\mathbb{E}[Y_n^i + Z_n^i] = \mathbb{E}[X_n^i]$$
which yields the assertion by theorem 4.1.1 and lemma 4.2.3. □

Lemma 4.2.8. *The choice (4.12) and $\eta_n^{i,2} = \eta_n^i - \eta_n^{i,1}$ yield for both $i \in \Sigma$*
$$\mathrm{Var}\left(Z_n^i\right) = \mathrm{O}(n).$$

Proof. The proof is done in three steps. In the first step, an upper bound $\mathrm{Var}(Z_n^i) = \mathrm{O}(n \log n)$ is derived by the transfer lemma 3.1.1. Afterwards, this bound is improved for the *Poissonized* variance $\mathrm{Var}(Z_N^i)$ with N Poisson distributed. Finally, the result is transferred to $\mathrm{Var}(Z_n^i)$ by the *Depoissonization* lemma 3.2.3.

To this end, first note that (4.9) implies for all $n \geq 1$ and $i \in \Sigma$
$$Z_{n+d}^i \stackrel{d}{=} Z_{I_n^i}^0 + Z_{n-I_n^i}^1 + \eta_n^{i,2} \tag{4.16}$$
where $(Z_n^0)_{n \geq 0}$, $(Z_n^1)_{n \geq 0}$ and $(I_n^i, \eta_n^{i,2})_{n \geq 0, i \in \Sigma}$ are independent. Now let
$$\nu_Z^i(n) := \mathbb{E}[Z_n^i], \qquad V_Z^i(n) := \mathrm{Var}(Z_n^i) \quad n \in \mathbb{N}_0, \; i \in \Sigma.$$
Then, (4.16) and lemma 3.0.4 yield
$$V_Z^i(n+d) = \mathbb{E}[V_Z^0(I_n^i)] + \mathbb{E}[V_Z^1(n - I_n^i)] + \mathrm{Var}(\nu_Z^0(I_n^i) + \nu_Z^1(n - I_n^i) + \eta_n^{i,2}). \tag{4.17}$$
Since ν_Z^0 and ν_Z^1 are Lipschitz-continuous (lemma 4.2.7), lemma 3.3.8 and the fact that I_n^i and $n - I_n^i$ follow the binomial distribution imply
$$\mathrm{Var}(\nu_Z^0(I_n^i)) = \mathrm{O}(n), \quad \mathrm{Var}(\nu_Z^1(n - I_n^i)) = \mathrm{O}(n), \quad i \in \Sigma.$$
Moreover, lemma 4.2.6 yields $\mathrm{Var}(\eta_n^{i,2}) = \mathrm{O}(n^{1-\varepsilon})$ and therefore, combined with lemma 4.2.2,
$$\mathrm{Var}(\nu_Z^0(I_n^i) + \nu_Z^1(n - I_n^i) + \eta_n^{i,2}) = \mathrm{O}(n).$$
Hence, (4.17) and the transfer lemma 3.1.1 imply that
$$V_Z^i(n) = \mathrm{O}(n \log n), \qquad i \in \Sigma. \tag{4.18}$$
In order to refine this bound, let N_λ be a Poisson $\Pi(\lambda)$ distributed random variable that is independent of $(Z_n^i, I_n^i, \eta_n^{i,2})_{n \geq 0, i \in \Sigma}$. Then, (4.16) implies
$$Z_{N_\lambda + d}^i \stackrel{d}{=} Z_{N_{\lambda p_{i0}}}^0 + Z_{M_{\lambda p_{i1}}}^1 + \eta_{N_\lambda}^{i,2} \tag{4.19}$$
where $N_{\lambda p_{i0}} := I_{N_\lambda}^i$ and $M_{\lambda p_{i1}} := N_\lambda - N_{\lambda p_{i0}}$. It is a well known fact, e.g. from *Poisson Processes* (marking each point with probability p_{i0}), and easy to compute that $N_{\lambda p_{i0}}$ and $M_{\lambda p_{i1}}$ are independent, $N_{\lambda p_{i0}}$ is Poisson $\Pi(\lambda p_{i0})$ distributed and $M_{\lambda p_{i1}}$ is Poisson $\Pi(\lambda p_{i1})$ distributed.

Now let $\widetilde{V}_i : \mathbb{R}^+ \to \mathbb{R}_0^+$, $i \in \Sigma$, be the variances after Poissonization defined as
$$\widetilde{V}_i(\lambda) := \mathrm{Var}(Z_{N_\lambda}^i).$$
The next step in the proof is to show the upper bound $\widetilde{V}_i(\lambda) = \mathrm{O}(\lambda)$ for both $i \in \Sigma$ (as $\lambda \to \infty$).

Such an upper bound follows by the transfer lemma 3.2.2 and the following recursive system which is shown next:

$$\begin{aligned}\widetilde{V}_0(\lambda) &= \widetilde{V}_0(\lambda p_{00}) + \widetilde{V}_1(\lambda p_{01}) + \mathrm{O}\left(\lambda^{\max\{3/4, 1-\varepsilon/2\}}\sqrt{\log \lambda}\right) & (\lambda \to \infty), \\ \widetilde{V}_1(\lambda) &= \widetilde{V}_0(\lambda p_{10}) + \widetilde{V}_1(\lambda p_{11}) + \mathrm{O}\left(\lambda^{\max\{3/4, 1-\varepsilon/2\}}\sqrt{\log \lambda}\right) & (\lambda \to \infty).\end{aligned} \quad (4.20)$$

First note that the bound (4.18) implies a similar upper bound on $\widetilde{V}_i(\lambda)$: A decomposition of the variance reveals

$$\widetilde{V}_i(\lambda) = \mathbb{E}[(Z^i_{N_\lambda} - \nu^i_Z(N_\lambda) + \nu^i_Z(N_\lambda) - \mathbb{E}[Z^i_{N_\lambda}])^2] = \mathbb{E}[V^i_Z(N_\lambda)] + \mathrm{Var}(\nu^i_Z(N_\lambda))$$

where the last equality holds since N_λ is independent of $(Z^i_n)_{n \geq 0}$ and

$$\mathbb{E}[Z^i_{N_\lambda}] = \mathbb{E}[\mathbb{E}[Z^i_{N_\lambda}|N_\lambda]] = \mathbb{E}[\nu^i_Z(N_\lambda)].$$

The Lipschitz-continuity of ν^i_Z (lemma 4.2.7) and lemma 3.3.8 yield, as $\lambda \to \infty$,

$$\mathrm{Var}(\nu^i_Z(N_\lambda)) = \mathrm{O}(\mathrm{Var}(N_\lambda)) = \mathrm{O}(\lambda).$$

Moreover, the rough upper bound $V^i_Z(n) = \mathrm{O}(n \log n)$ implies, as $\lambda \to \infty$,

$$\mathbb{E}[V^i_Z(N_\lambda)] = \mathrm{O}(\mathbb{E}[N_\lambda \log N_\lambda]) = \mathrm{O}(\lambda \log \lambda)$$

where the second bound follows by an easy calculation on the Poisson distribution that can be found in the appendix, lemma A.2.3. Therefore, the *Poissonized* variance is bounded by

$$\widetilde{V}_i(\lambda) = \mathrm{O}(\lambda \log \lambda), \quad i \in \Sigma, \quad (\lambda \to \infty) \quad (4.21)$$

and, in particular, $\sup_{\lambda \in (0,a]} \widetilde{V}_i(\lambda) < \infty$ for all $a \in \mathbb{R}^+$.

Furthermore, lemma 4.2.6 implies

$$\begin{aligned}\mathrm{Var}(\eta^{i,2}_{N_\lambda}) &= \mathbb{E}[\mathbb{E}[(\eta^{i,2}_{N_\lambda} - \mathbb{E}[\eta^{i,2}_{N_\lambda}|N_\lambda])^2|N_\lambda]] + \mathrm{Var}(\mathbb{E}[\eta^{i,2}_{N_\lambda}|N_\lambda]) \\ &= \mathrm{O}(\mathbb{E}[N^{1-\varepsilon}_\lambda]) = \mathrm{O}(\lambda^{1-\varepsilon})\end{aligned} \quad (4.22)$$

where $\mathbb{E}[N^{1-\varepsilon}_\lambda] = \mathrm{O}(\lambda^{1-\varepsilon})$ is not hard to compute and can be found in the appendix, lemma A.2.3.

The distributional equation (4.19) and the upper bounds (4.21) and (4.22) yield by lemma 4.2.2 that

$$\begin{aligned}\mathrm{Var}(Z^i_{N_\lambda+d}) &= \mathrm{Var}(Z^0_{N_{\lambda p_{i0}}} + Z^1_{M_{\lambda p_{i1}}} + \eta^{i,2}_{N_\lambda}) \\ &= \mathrm{Var}(Z^0_{N_{\lambda p_{i0}}} + Z^1_{M_{\lambda p_{i1}}}) + \mathrm{O}\left(\sqrt{\lambda^{2-\varepsilon} \log \lambda}\right) \\ &= \widetilde{V}_0(\lambda p_{i0}) + \widetilde{V}_1(\lambda p_{i1}) + \mathrm{O}\left(\sqrt{\lambda^{2-\varepsilon} \log \lambda}\right)\end{aligned}$$

where the last equality holds because $Z^0_{N_{\lambda p_{i0}}}$ and $Z^1_{M_{\lambda p_{i1}}}$ are independent.

Finally, it only remains to show that $\mathrm{Var}(Z^i_{N_\lambda+d}) = \widetilde{V}_i(\lambda) + \mathrm{O}(\lambda^{3/4})$ in order to get (4.20). By the identity $\mathrm{Var}(Z^i_{N_\lambda+d}) = \mathbb{E}[V^i_Z(N_\lambda+d)] + \mathrm{Var}(\nu^i_Z(N_\lambda+d))$ it is sufficient to show that

4.2. ANALYSIS OF THE VARIANCE

(a) $\mathbb{E}[V_Z^i(N_\lambda + d)] = \mathbb{E}[V_Z^i(N_\lambda)] + O(\lambda^{3/4})$,

(b) $\text{Var}(\nu_Z^i(N_\lambda + d)) = \text{Var}(\nu_Z^i(N_\lambda)) + O(\sqrt{\lambda})$.

Note that (a) follows from the rough upper bound $V_Z^i(n) = O(n \log n)$ and lemma 3.2.1 and (b) from the Lipschitz-continuity of ν_Z^i (lemma 4.2.7) since the Lipschitz-continuity implies

$$\text{Var}(\nu_Z^i(N_\lambda)) = O(\text{Var}(N_\lambda)), \qquad \text{Var}(\nu_Z^i(N_\lambda + d) - \nu_Z^i(N_\lambda)) = O(1)$$

and therefore, by lemma 4.2.2,

$$\text{Var}(\nu_Z^i(N_\lambda + d)) = \text{Var}(\nu_Z^i(N_\lambda) + \nu_Z^i(N_\lambda + d) - \nu_Z^i(N_\lambda)) = \text{Var}(\nu_Z^i(N_\lambda)) + O(\sqrt{\lambda}).$$

Combined with the previous observations, this yields the recursive system (4.20). Lemma 3.2.2 implies for such a system that

$$\widetilde{V}_i(\lambda) = O(\lambda), \quad i \in \Sigma \qquad (\lambda \to \infty).$$

The last step of the proof is to *Depoissonize* this result, i.e. to transfer the upper bound on $\widetilde{V}_i(\lambda)$ into the upper bound $V_Z^i(n) = O(n)$.

Recall that $\widetilde{V}_i(\lambda) = \mathbb{E}[V_Z^i(N_\lambda)] + \text{Var}(\nu_Z^i(N_\lambda))$ and that the Lipschitz-continuity of ν_Z^i yields

$$\text{Var}(\nu_Z^i(N_\lambda)) = O(\text{Var}(N_\lambda)) = O(\lambda).$$

Hence, the upper bound on $\widetilde{V}_i(\lambda)$ implies, as $\lambda \to \infty$,

$$\mathbb{E}[V_Z^i(N_\lambda)] = \widetilde{V}_i(\lambda) - \text{Var}(\nu_Z^i(N_\lambda)) = O(\lambda), \quad i \in \Sigma. \tag{4.23}$$

The assertion follows from the *Depoissonization* lemma 3.2.3 and the following bound on the increments which is shown next:

$$\Delta V_Z^i(n) = O(\sqrt{n}).$$

To this end, note that (4.17) and lemma 4.1.3 imply

$$\Delta V_Z^i(n+d) = p_{i0}\mathbb{E}[\Delta V_Z^0(I_n^i)] + p_{i1}\mathbb{E}[\Delta V_Z^1(n - I_n^i)] + \Delta\text{Var}(\nu_Z^0(I_n^i) + \nu_Z^1(n - I_n^i) + \eta_n^{i,2}).$$

Hence, by lemma 3.1.3 it is sufficient to show that

$$\Delta\text{Var}(\nu_Z^0(I_n^i) + \nu_Z^1(n - I_n^i) + \eta_n^{i,2}) = O(\sqrt{n}). \tag{4.24}$$

The Lipschitz-continuity of ν_Z^i, corollary 4.2.6 and lemma 4.2.2 imply

$$\text{Var}(\nu_Z^0(I_n^i) + \nu_Z^1(n - I_n^i) + \eta_n^{i,2}) = O(n), \qquad \text{Var}(\Delta\nu_Z^0(I_n^i) + \Delta\nu_Z^1(n - I_n^i) + \Delta\eta_n^{i,2}) = O(1).$$

Therefore, lemma 4.2.2 yields

$$\begin{aligned}&\text{Var}(\nu_Z^0(I_{n+1}^i) + \nu_Z^1(n+1 - I_{n+1}^i) + \eta_{n+1}^{i,2}) \\ &= \text{Var}(\nu_Z^0(I_n^i) + \nu_Z^1(n - I_n^i) + \eta_n^{i,2} + \Delta\nu_Z^0(I_n^i) + \Delta\nu_Z^1(n - I_n^i) + \Delta\eta_n^{i,2}) \\ &= \text{Var}(\nu_Z^0(I_n^i) + \nu_Z^1(n - I_n^i) + \eta_n^{i,2}) + O(\sqrt{n})\end{aligned}$$

which is (4.24). Hence, the *Depoissonization* lemma 3.2.3 applied to (4.23) yields that

$$|V_Z^i(n) - \mathbb{E}[V_Z^i(N_n)]| = O(n).$$

Finally, the bound $\mathbb{E}[V_Z^i(N_n)] = O(n)$ given in (4.23) implies the assertion. $\square$

Proof of theorem 4.2.1. Let $(Y_n^i, Z_n^i)_{n \geq 0, i \in \Sigma}$ be chosen such that (4.8) and (4.9) hold with toll terms given by (4.12). Consider the sum

$$S_n^i := Y_n^i + Z_n^i, \quad i \in \Sigma, n \in \mathbb{N}_0.$$

Note that (4.8) and (4.9) imply for both $i \in \Sigma$ that $S_n^i = 0$ for all $n \leq \max\{d, 1\}$ and

$$S_{n+d}^i \stackrel{d}{=} S_{I_n^i}^0 + S_{n-I_n^i}^1 + \eta_n^i, \quad n \geq 2 - \min\{d, 1\}.$$

It is not hard to check that this implies $\mathbb{E}[S_n^i] = \mathbb{E}[X_n^i]$ and $\text{Var}(S_n^i) = \text{Var}(X_n^i)$ which is discussed in detail in the second part of lemma 4.2.9 at the end of the section.

Hence, the lemmata 4.2.4, 4.2.8 and 4.2.2 yield of both $i \in \Sigma$, as $n \to \infty$,

$$\text{Var}(X_n^i) = \text{Var}(Y_n^i + Z_n^i) = \sigma^2 n \log n + O\left(n \sqrt{\log n}\right)$$

which is the assertion. □

We finish the analysis of the variances with the missing proof for the existence of $(Y_n^i, Z_n^i)_{n \geq 0}$:

Lemma 4.2.9. *There exist sequences $(Y_n^0, Z_n^0)_{n \geq 0}$ and $(Y_n^1, Z_n^1)_{n \geq 0}$ of pairs of real valued random variables with finite second moments that satisfy the initial conditions (4.8) and the system (4.9) of distributional equations.*

Moreover, suppose that $(X_n^0)_{n \geq 0}$ and $(X_n^1)_{n \geq 0}$ are sequences of real valued random variables with finite second moments that satisfy the initial conditions $X_n^0 = X_n^1 = 0$ for $n \leq \max\{d, 1\}$ and the distributional recursions (4.1). Then, for all $n \in \mathbb{N}_0$ and $i \in \Sigma$,

$$\mathbb{E}[X_n^i] = \mathbb{E}[Y_n^i + Z_n^i], \quad \text{Var}(X_n^i) = \text{Var}(Y_n^i + Z_n^i).$$

Proof. First note that in the case $d \geq 1$ the existence of proper pairs of random variables is trivial because $(Y_n^i, Z_n^i), i \in \Sigma$, may be recursively defined as

$$\begin{pmatrix} Y_n^i \\ Z_n^i \end{pmatrix} := \begin{pmatrix} 0 \\ 0 \end{pmatrix}, \quad n \leq d$$

and, for all $n \geq d + 1$

$$\mathcal{L}\left(\begin{pmatrix} Y_n^i \\ Z_n^i \end{pmatrix}\right) := \mathcal{L}\left(\begin{pmatrix} Y_{I_{n-d}^i}^0 \\ Z_{I_{n-d}^i}^0 \end{pmatrix} + \begin{pmatrix} Y_{n-d-I_{n-d}^i}^1 \\ Z_{n-d-I_{n-d}^i}^1 \end{pmatrix} + \begin{pmatrix} \eta_{n-d}^{i,1} \\ \eta_{n-d}^{i,2} \end{pmatrix}\right).$$

with $(Y_0^0, Z_0^0, \ldots, Y_{n-d}^0, Z_{n-d}^0)$, $(Y_0^1, Z_0^1, \ldots, Y_{n-d}^1, Z_{n-d}^1)$ and (I_n^i, η_n^i) independent. Recall that, for all $n \in \mathbb{N}_0$ and $i \in \Sigma$, $\eta_n^{i,1}$ is some constant defined in (4.12) and that $\eta_n^i = g_n^i(I_n^i)$ for some function $g_n^i : \{0, \ldots, n\} \to \mathbb{R}$. Therefore, $|\eta_n^i| \leq C_n^i$ for some constant given by $C_n^i = \max\{g_n^i(m) : 0 \leq m \leq n\}$. This implies that $\eta_n^{i,2} = \eta_n^i - \eta_n^{i,1}$ is also bounded which implies by induction on n that Z_n^i and Y_n^i are bounded (by a constant that depends on n). In particular, Y_n^i and Z_n^i have finite moments of any order for all $n \in \mathbb{N}_0$ and $i \in \Sigma$ which finishes the proof of the existence for $d \geq 1$.

The main idea of the proof for $d = 0$ is to define (Y_n^i, Z_n^i) for $n \geq 2$ as the series that appears when the right hand side of (4.9) is iterated infinitely many times. To this end, some notation is required:

4.2. ANALYSIS OF THE VARIANCE

For $n \in \mathbb{N}_0$ and $i \in \Sigma$ let $I_i(n) := \sum_{j=1}^n L_j^i$ where $(L_j^i)_{j \in \mathbb{N}}$ is a sequence of independent Bernoulli $B(p_{i0})$ distributed random variables. Moreover, for every $k \geq 1$ and $J \in \{0,1\}^k$ let $(I_{i,0}^J(n), I_{i,1}^J(n))_{n \geq 0}$ be an independent copy of $(I_i(n), n - I_i(n))_{n \geq 0}$. Then, for any $k \geq 1$ and $J = (j_1, \ldots, j_k) \in \{0,1\}^k$, recursively define random variables $\mathfrak{I}_i^J(n)$ as

$$\mathfrak{I}_i^J(n) := \begin{cases} I_i(n), & \text{if } k = 1 \text{ and } j_1 = 0, \\ n - I_i(n), & \text{if } k = 1 \text{ and } j_1 = 1, \\ I_{j_{k-1},j_k}^{(j_1,\ldots,j_{k-1})} \left(\mathfrak{I}_i^{(j_1,\ldots,j_{k-1})}(n) \right), & \text{otherwise.} \end{cases}$$

In the context of *Radix Sort*, these random variables can be interpreted as follows: consider a set $\mathcal{X}_n^i = \{\Xi_1^{(i)}, \ldots, \Xi_n^{(i)}\}$ of n independent and identically distributed random strings where each string is generated by a *Markov Source* with initial distribution $\mu = p_{i0}\delta_0 + p_{i1}\delta_1$ and transition matrix $(p_{kl})_{k,l \in \{0,1\}}$ (see definition 1.2.2). Then, $\mathfrak{I}_i^J(n)$ with $J = (j_1, \ldots, j_k)$ denotes the number of strings that start with prefix J, i.e.

$$\mathfrak{I}_i^J(n) \stackrel{d}{=} |\{(\xi_j)_{j \geq 1} \in \mathcal{X}_n^i : (\xi_1, \ldots, \xi_k) = J\}|. \tag{4.25}$$

Now let $\tilde{g}_n^i : \{0, \ldots, n\} \to \mathbb{R}^2$ be defined as

$$\tilde{g}_n^i(m) = \begin{pmatrix} \eta_n^{i,1} \\ g_n^i(m) - \eta_n^{i,1} \end{pmatrix}.$$

Since $\eta_n^i = g_n^i(I_n^i)$, this yields for $n \geq 2$ and $i \in \Sigma$

$$\begin{pmatrix} \eta_n^{i,1} \\ \eta_n^{i,2} \end{pmatrix} = \tilde{g}_n^i(I_n^i).$$

With the convention $\tilde{g}_0^i(0) = \tilde{g}_1^i(0) = \tilde{g}_1^i(1) = (0,0)^T$, let the random variables (Y_n^i, Z_n^i) for $n \geq 2$ and $i \in \Sigma$ be defined as

$$\begin{pmatrix} Y_n^i \\ Z_n^i \end{pmatrix} := \tilde{g}_n^i \left(\mathfrak{I}_i^{(0)}(n) \right) + \sum_{k=1}^{\infty} \sum_{\substack{J \in \{0,1\}^k \\ (j_1, \ldots, j_k) := J}} \tilde{g}_{\mathfrak{I}_i^J(n)}^{j_k} \left(\mathfrak{I}_i^{(j_1, \ldots, j_k, 0)}(n) \right). \tag{4.26}$$

Note that (4.26) also holds for $n \leq 1$ since in this case all summands are zero. The following properties need to be shown for the random variables defined in (4.26) in order to finish the proof of the existence ($d = 0$):

(a) Almost surely, a (random) integer N exists such that

$$\tilde{g}_{\mathfrak{I}_i^J(n)}^{j_k} \left(\mathfrak{I}_i^{(j_1,\ldots,j_k,0)}(n) \right) = 0, \quad J = (j_1, \ldots, j_k) \in \{0,1\}^k, \quad k \geq N. \tag{4.27}$$

In particular, (Y_n^i, Z_n^i) is well-defined and almost surely finite.

(b) The definition (4.26) yields random variables $(Y_n^0, Z_n^0)_{n \geq 0}$ and $(Y_n^1, Z_n^1)_{n \geq 0}$ that satisfy the system (4.9) of distributional recursions.

(c) The constructed random variables have $\mathbb{E}[(Y_n^i)^2] < \infty$ and $\mathbb{E}[(Z_n^i)^2] < \infty$ for all $n \in \mathbb{N}$ and $i \in \Sigma$.

For part (a) recall that $\tilde{g}_n^i = (0,0)^T$ for $n \leq 1$. Moreover, note that $\mathfrak{I}_i^J(n)$ is decreasing for increasing J in the sense that $J = (j_1, \ldots, j_k)$ and $J' = (j_1, \ldots, j_k, j'_{k+1}, \ldots, j'_{k+l})$ for some $k, l \in \mathbb{N}$ implies $\mathfrak{I}_i^J(n) \geq \mathfrak{I}_i^{J'}(n)$.

Hence, (4.27) holds for all $k \geq N$ with

$$N = \min\left\{\ell \geq 1 : \mathfrak{I}_i^J(n) \leq 1 \text{ for all } J \in \{0,1\}^\ell\right\}.$$

It only remains to show that $N < \infty$ almost surely holds for N defined as above. Connected to *Radix Sort*, $N = \infty$ implies that *Radix Sort* does not terminate because a pair of strings with common ℓ-prefix exists for all $\ell \geq 1$. In particular,

$$\mathbb{P}(N = \infty) \leq \mathbb{P}(B_n^i = \infty)$$

where B_n^i denotes the number of bucket operations performed by radix sort under the Markov Source Model with initial distribution $p_{i0}\delta_0 + p_{i1}\delta_1$.

As already shown in the proof of corollary 2.1.1, B_n^i has a finite expectation and therefore, $\mathbb{P}(N = \infty) \leq \mathbb{P}(B_n^i = \infty) = 0$. Moreover, note that N is a lower bound on the number of recursive calls of *Radix Sort* (more precisely, N equals the height of the corresponding Trie, see section 2.2 for a definition) and therefore, N is bounded by B_n^i when coupled properly. In particular, the square integrability of B_n^i discussed in corollary 2.1.1 implies that $\mathbb{E}[N^2] < \infty$.

For part (b) let $(\mathfrak{I}_0^J(n))_{n \geq 0, J \in \bigcup_{k \geq 1}\{0,1\}^k}$, $(\mathfrak{I}_1^J(n))_{n \geq 0, J \in \bigcup_{k \geq 1}\{0,1\}^k}$ and $\tilde{I}_n^i$ be independent with $\tilde{I}_n^i$ following the binomial distribution $B(n, p_{i0})$. Moreover, for $k \geq 1$ and $J = (j_1, \ldots, j_k) \in \{0,1\}^k$, let $\tilde{\mathfrak{I}}_i^J(n)$ be defined as

$$\tilde{\mathfrak{I}}_i^J(n) := \begin{cases} \tilde{I}_n^i, & \text{if } k=1, j_1 = 0, \\ n - \tilde{I}_n^i, & \text{if } k=1, j_1 = 1, \\ \mathfrak{I}_0^{(j_2, \ldots, j_k)}(\tilde{I}_n^i), & \text{if } k \geq 2, j_1 = 0, \\ \mathfrak{I}_1^{(j_2, \ldots, j_k)}(n - \tilde{I}_n^i), & \text{if } k \geq 2, j_1 = 1. \end{cases}$$

Note that $(\tilde{\mathfrak{I}}_i^J(n))_{n \geq 0, J \in \bigcup_{k \geq 1}\{0,1\}^k}$ is distributed as $(\mathfrak{I}_i^J(n))_{n \geq 0, J \in \bigcup_{k \geq 1}\{0,1\}^k}$.

Hence, definition (4.26) yields

$$\begin{pmatrix} Y_{\tilde{I}_n^i}^0 \\ Z_{\tilde{I}_n^i}^0 \end{pmatrix} + \begin{pmatrix} Y_{n-\tilde{I}_n^i}^1 \\ Z_{n-\tilde{I}_n^i}^1 \end{pmatrix} + \tilde{g}_n^i(\tilde{I}_n^i) \stackrel{d}{=} \tilde{g}_{\tilde{\mathfrak{I}}_i^{(0)}(n)}^0\left(\tilde{\mathfrak{I}}_i^{(0,0)}(n)\right) + \sum_{k=1}^{\infty} \sum_{\substack{J \in \{0,1\}^k \\ (j_1, \ldots, j_k) := J}} \tilde{g}_{\tilde{\mathfrak{I}}_i^{(0,j_1,\ldots,j_k)}(n)}^{j_k}\left(\tilde{\mathfrak{I}}_i^{(0,j_1,\ldots,j_k,0)}(n)\right)$$

$$+ \tilde{g}_{\tilde{\mathfrak{I}}_i^{(1)}(n)}^1\left(\tilde{\mathfrak{I}}_i^{(1,0)}(n)\right) + \sum_{k=1}^{\infty} \sum_{\substack{J \in \{0,1\}^k \\ (j_1, \ldots, j_k) := J}} \tilde{g}_{\tilde{\mathfrak{I}}_i^{(1,j_1,\ldots,j_k)}(n)}^{j_k}\left(\tilde{\mathfrak{I}}_i^{(1,j_1,\ldots,j_k,0)}(n)\right)$$

$$+ \tilde{g}_n^i\left(\tilde{\mathfrak{I}}_i^{(0)}(n)\right).$$

$$= \tilde{g}_n^i\left(\tilde{\mathfrak{I}}_i^{(0)}(n)\right) + \sum_{k=1}^{\infty} \sum_{\substack{J \in \{0,1\}^k \\ (j_1, \ldots, j_k) := J}} \tilde{g}_{\tilde{\mathfrak{I}}_i^{(j_1,\ldots,j_k)}(n)}^{j_k}\left(\tilde{\mathfrak{I}}_i^{(j_1,\ldots,j_k,0)}(n)\right)$$

$$\stackrel{d}{=} \begin{pmatrix} Y_n^i \\ Z_n^i \end{pmatrix}$$

4.2. ANALYSIS OF THE VARIANCE

where the rearrangement of the series in the second equality is justified by part (a) and the last equality in distribution holds by definition (4.26). Hence, $(Y_n^0, Z_n^0)_{n \geq 0}$ and $(Y_n^1, Z_n^1)_{n \geq 0}$ satisfy the system (4.9) of distributional recursions.

For part (c) note that definition (4.12) and lemma 4.2.5 yield for both $i \in \Sigma$ that $\eta_n^{i,1} = 0$ for $n \leq 1$ and
$$|\eta_n^{i,1}| \leq Cn, \qquad n \geq 2,$$
with a suitable constant $C > 0$. Let N be chosen as in (a). Then,

$$|Y_n^i| \leq Cn + \sum_{i=1}^{N} \sum_{\substack{J \in \{0,1\}^k \\ (j_1,\ldots,j_k) := J}} C\mathfrak{I}_i^J(n)$$

$$= Cn + \sum_{i=1}^{N} Cn$$

$$= C(N+1)n$$

where the second equality holds because $(\mathfrak{I}_i^J(n))_{J \in \{0,1\}^k}$ follows the multinomial distribution for any $k \geq 1$. Therefore, the integrability of Y_n^i is implied by the integrability of N which was discussed in part (a).

The integrability of Z_i is shown by the same arguments and $|\eta_n^{i,2}| \leq |\eta_n^i| + |\eta_n^{i,1}|$ which has a linear upper bound by the previous observation on $\eta_n^{i,1}$ and assumption (4.2) on η_n^i.

Now let $S_n^i := Y_n^i + Z_n^i$ for $n \in \mathbb{N}_0$ and $i \in \Sigma$. It remains to show that S_n^i and X_n^i have the same mean and variance. Note that the initial conditions (4.8) and taking the sum in (4.9) imply that $S_n^i = 0$ for $n \leq d$ and that for $n \geq 1$

$$\begin{aligned} S_{n\mid d}^0 &\stackrel{d}{=} S_{I_n^0}^0 + S_{n\mid I_n^0}^1 + \eta_n^0, \\ S_{n+d}^1 &\stackrel{d}{=} S_{I_n^1}^0 + S_{n-I_n^1}^1 + \eta_n^1, \end{aligned} \qquad (4.28)$$

with $(S_n^0)_{n \geq 0}$, $(S_n^1)_{n \geq 0}$ and (I_n^0, I_n^1) being independent. Hence, $(S_n^i)_{n \geq 0}$, $i \in \Sigma$ satisfies the same initial conditions and the same system of distributional equations as $(X_n^i)_{n \geq 0}$, $i \in \Sigma$. It only remains to show that these equations uniquely define the first two moments of $(X_n^i)_{n \geq 0}$, $i \in \Sigma$. For $d \geq 1$ the initial conditions and distributional equations uniquely define the distribution of X_n^i for all $n \in \mathbb{N}_0$ and $i \in \Sigma$ which implies the equality of all moments in this case.

For $d = 0$ note that
$$\mathbb{E}[X_n^i] = \sum_{k=0}^{n} \mathbb{P}(I_n^i = k)\bigg(\mathbb{E}[X_k^0 + X_{n-k}^1] + g_n(k)\bigg), \qquad n \geq 2,$$

and therefore
$$(1 - p_{00}^n)\mathbb{E}[X_n^0] - p_{01}^n\mathbb{E}[X_n^1] = \sum_{k=1}^{n-1} \mathbb{P}(I_n^0 = k)\bigg(\mathbb{E}[X_k^0 + X_{n-k}^1] + g_n^0(k)\bigg),$$

$$-p_{10}^n\mathbb{E}[X_n^0] + (1 - p_{11}^n)\mathbb{E}[X_n^1] = \sum_{k=1}^{n-1} \mathbb{P}(I_n^1 = k)\bigg(\mathbb{E}[X_k^0 + X_{n-k}^1] + g_n^1(k)\bigg).$$

Given $(\mathbb{E}[X_k^0], \mathbb{E}[X_k^1])_{0 \le k \le n-1}$, this system has a unique solution for $(\mathbb{E}[X_n^0], \mathbb{E}[X_n^1])$ because, for any $p_{00}, p_{11} \in (0,1)$,

$$\det\left(\begin{pmatrix} 1 - p_{00}^n & -p_{01}^n \\ -p_{10}^n & 1 - p_{11}^n \end{pmatrix}\right) = \underbrace{(1 - p_{00}^n)}_{>(1-p_{00})^n}\underbrace{(1 - p_{11}^n)}_{>(1-p_{11})^n} - p_{01}^n p_{10}^n > 0.$$

Hence, the initial conditions and (4.1) uniquely define the first moment of X_n^0 and X_n^1 for $n \in \mathbb{N}_0$. Similar arguments also work for the second moments: the independence of $(X_n^0)_{n \ge 0}$, $(X_n^1)_{n \ge 0}$ and I_n^i in (4.1) yields for both $i \in \Sigma$

$$\mathbb{E}[(X_n^i)^2] = \sum_{k=0}^n \mathbb{P}(I_n^i = k)\mathbb{E}[(X_k^0 + X_{n-k}^1 + g_n^i(k))^2]$$

$$= \sum_{k=0}^n \mathbb{P}(I_n^i = k)\bigg(\mathbb{E}[(X_k^0)^2] + \mathbb{E}[(X_k^1)^2] + 2\mathbb{E}[X_k^0]\mathbb{E}[X_{n-k}^1]$$

$$+ (g_n^i(k))^2 + 2g_n^i(k)\mathbb{E}[X_k^0] + 2g_n^i(k)\mathbb{E}[X_{n-k}^1]\bigg)$$

which, given $(\mathbb{E}[(X_k^0)^2], \mathbb{E}[(X_k^1)^2])_{0 \le k \le n-1}$ and $(\mathbb{E}[X_k^0], \mathbb{E}[X_k^1])_{0 \le k \le n}$, has a unique solution for $(\mathbb{E}[(X_n^0)^2], \mathbb{E}[(X_n^1)^2])$.

Therefore, the initial conditions and (4.1) uniquely define the first two moments of X_n^0 and X_n^1 which yields the second part of the assertion by (4.28) and the initial conditions therein. □

4.3 Limit Theorems (Contraction Method)

The previous results on mean and variance enable an application of the *Contraction Method* presented in section 3.3. Recall that $(X_n^0)_{n \ge 0}$ and $(X_n^1)_{n \ge 0}$ satisfy the system (4.1) of distributional recursions with *toll terms* that satisfy the conditions (4.2). Moreover, recall that X_n^0 and X_n^1 have a finite s-th moment for all $n \in \mathbb{N}_0$ and some $s \in (2, 3]$.

The transition matrix $P = (p_{ij})_{i,j \in \Sigma}$ of the *Markov Source* satisfies the conditions (1.9):

$$p_{ij} \in (0,1) \text{ for all } (i,j) \in \Sigma^2, \qquad p_{ij} \ne \frac{1}{2} \text{ for some } (i,j) \in \Sigma^2.$$

Throughout the section, mean and variance of X_n^i are abbreviated by

$$\sigma_i(n) := \sqrt{\text{Var}(X_n^i)}, \qquad \nu_i(n) := \mathbb{E}[X_n^i], \quad n \in \mathbb{N}_0, i \in \Sigma.$$

As mentioned in section 3.3, an application of the contraction method in this context requires an exact normalization. To this end, let $(Y_n^i)_{n \ge 0}$, $i \in \Sigma$, be defined as

$$Y_n^i := \begin{cases} 0, & \text{if } \sigma_i(n) = 0, \\ \frac{X_n^i - \nu_i(n)}{\sigma_i(n)}, & \text{if } \sigma_i(n) > 0. \end{cases} \qquad (4.29)$$

The main result in this section is the following limit theorem:

4.3. LIMIT THEOREMS (CONTRACTION METHOD)

Theorem 4.3.1. *Let $(X_n^0)_{n\geq 0}$ and $(X_n^1)_{n\geq 0}$ be sequences of real valued random variables with finite s-th moments for some $s \in (2,3]$ that satisfy the initial conditions $X_n^0 = X_n^1 = 0$ for $n \leq \max\{d,1\}$ and the distributional recursions (4.1). Assume that the toll functions η_n^0 and η_n^1 in (4.1) satisfy the conditions (4.2).*

Then, the normalized random variables defined in (4.29) satisfy for both $i \in \Sigma$, as $n \to \infty$,

$$Y_n^i \xrightarrow{d} \mathcal{N}(0,1).$$

Here, $\mathcal{N}(0,1)$ denotes a random variable with the standard normal distribution.

Note that the system (4.1) yields a similar system for Y_n^0 and Y_n^1 after rescaling:

$$\begin{aligned} Y_{n+d}^0 &\stackrel{d}{=} \frac{\sigma_0(I_n^0)}{\sigma_0(n+d)} Y_{I_n^0}^0 + \frac{\sigma_1(n-I_n^0)}{\sigma_0(n+d)} Y_{n-I_n^0}^1 + b_0(n), \quad n \in \{m \in \mathbb{N} : \sigma_0(m+d) > 0\}, \\ Y_{n+d}^1 &\stackrel{d}{=} \frac{\sigma_0(I_n^1)}{\sigma_1(n+d)} Y_{I_n^1}^0 + \frac{\sigma_1(n-I_n^1)}{\sigma_1(n+d)} Y_{n-I_n^1}^1 + b_1(n), \quad n \in \{m \in \mathbb{N} : \sigma_1(m+d) > 0\}, \end{aligned} \quad (4.30)$$

where $(Y_n^0)_{n\geq 0}$, $(Y_n^1)_{n\geq 0}$ and $(I_n^i, b_i(n))_{i \in \Sigma}$ are independent, I_n^i follows the binomial distribution $B(n, p_{i0})$ and $b_i(n)$ is given by

$$b_i(n) = \frac{\nu_0(I_n^i) + \nu_1(n - I_n^i) + \eta_n^i - \nu_i(n+d)}{\sigma_i(n+d)}.$$

The asymptotic behavior of mean and variance and the assumption on η_n^i lead to the following limits for the coefficients:

Lemma 4.3.2. *The coefficients appearing in (4.30) satisfy for both $i \in \Sigma$, as $n \to \infty$,*

$$\frac{\upsilon_0(I_n^i)}{\sigma_i(n+d)} \to \sqrt{p_{i0}} \quad \text{a.s. and in } L_p \text{ for any } p \geq 1,$$

$$\frac{\sigma_1(n - I_n^i)}{\sigma_i(n+d)} \to \sqrt{p_{i1}} \quad \text{a.s. and in } L_p \text{ for any } p \geq 1,$$

$$\|b_i(n)\|_3 \to 0.$$

Proof. Recall that $\sigma_i(n) \sim \sigma\sqrt{n \log n}$ by theorem 4.2.1. The identity

$$\frac{\log m}{\log n} = 1 + \frac{\log m - \log n}{\log n} = 1 + \frac{\log(m/n)}{\log n}, \quad m, n \in \mathbb{N}$$

yields almost surely, as $n \to \infty$,

$$\frac{\sigma_0(I_n^i)}{\sigma_i(n+d)} = \sqrt{\frac{I_n^i}{n}\left(1 + \frac{\log(I_n^i/n)}{\log n}\right)}(1 + o(1)).$$

The strong law of large numbers implies that almost surely, as $n \to \infty$,

$$\frac{\sigma_0(I_n^i)}{\sigma_i(n+d)} \to \sqrt{p_{i0}}.$$

This result may also be transferred to L_p-convergence by the dominated convergence theorem: note that $\sigma_i(n) \sim \sigma\sqrt{n \log n}$ implies the existence of constants $C_1, C_2 > 0$ and $n_1 \in \mathbb{N}$ such that

$$\sigma_i(n) \leq C_1\sqrt{(n+1)\log(n+1)} \quad \text{for all } n \in \mathbb{N}_0,\, i \in \Sigma,$$
$$\sigma_i(n) \geq C_2\sqrt{(n+1)\log(n+1)} \quad \text{for all } n \geq n_1,\, i \in \Sigma,$$

which yields that $\sigma_0(I_n^i)/\sigma_i(n)$ is bounded by C_1/C_2 for all $n \geq n_1$, $i \in \Sigma$.

One obtains by the same arguments, as $n \to \infty$,

$$\frac{\sigma_1(n - I_n^i)}{\sigma_i(n+d)} \to \sqrt{p_{i1}} \quad \text{a.s. and in } L_p \text{ for any } p \geq 1.$$

It remains to show the L_3-convergence of $b_0(n)$ and $b_1(n)$. To this end, let $h : [0,1] \to \mathbb{R}$ be defined as

$$h(x) = \begin{cases} 0, & \text{if } x = 0, \\ x \log x, & \text{otherwise.} \end{cases}$$

Note that lemma 3.0.4 and theorem 4.1.1 yield

$$\|b_i(n)\|_3 = \frac{1}{\sigma_i(n+d)}\|\nu_0(I_n^i) - \mathbb{E}[\nu_0(I_n)] + \nu_1(n - I_n^i) - \mathbb{E}[\nu_1(n - I_n^i)] + \eta_n^i - \mathbb{E}[\eta_n^i]\|_3$$
$$\leq \frac{1}{H\sigma_i(n+d)}\|(h(I_n^i) + h(n - I_n^i)) - \mathbb{E}[h(I_n^i) + h(n - I_n^i)])\|_3$$
$$+ \frac{1}{\sigma_i(n+d)}\left(\|f_0(I_n^i) - \mathbb{E}[f_0(I_n^i)]\|_3 + \|f_1(n - I_n^i) - \mathbb{E}[f_1(n - I_n^i)]\|_3 + \|\eta_n^i - \mathbb{E}[\eta_n^i]\|_3\right)$$

where $f_i : \mathbb{N}_0 \to \mathbb{R}, n \mapsto \nu_i(n) - \frac{1}{H}n\log n$, $i \in \Sigma$, are Lipschitz-continuous functions. Hence, it only remains to show that, as $n \to \infty$

(a) $\frac{1}{H\sigma_i(n+d)}\|(h(I_n^i) + h(n - I_n^i)) - \mathbb{E}[h(I_n^i) + h(n - I_n^i)]\|_3 \to 0$,

(b) $\frac{1}{\sigma_i(n+d)}\|f_0(I_n^i) - \mathbb{E}[f_0(I_n^i)]\|_3 \to 0$, $\quad \frac{1}{\sigma_i(n+d)}\|f_1(n - I_n^i) - \mathbb{E}[f_1(n - I_n^i)]\|_3 \to 0$,

(c) $\frac{1}{\sigma_i(n+d)}\|\eta_n^i - \mathbb{E}[\eta_n^i]\|_3 \to 0$.

For part (a) note that (adding $n\log n - (I_n^i + n - I_n^i)\log n = 0$)

$$\|h(I_n^i) + h(n - I_n^i) - \mathbb{E}[h(I_n^i) + h(n - I_n^i)]\|_3$$
$$= \|nh(I_n^i/n) + nh(1 - I_n^i/n) - \mathbb{E}[nh(I_n^i/n) + nh(1 - I_n^i/n)]\|_3$$
$$\leq n\left(\|h(I_n^i/n) - \mathbb{E}[h(I_n^i/n)]\|_3 + \|h(1 - I_n^i/n) - \mathbb{E}[h(1 - I_n^i/n)]\|_3\right).$$

Since h is Lipschitz-continuous on $[\varepsilon, 1]$ for any $\varepsilon > 0$ and $I_n^i/n > \varepsilon$ with high probability for any $\varepsilon < p_{i0}$, it is not hard to check that, similar to lemma 3.3.8 on Lipschitz functions,

$$\|h(I_n^i/n) - \mathbb{E}[h(I_n^i/n)]\|_3 = O\left(\|I_n^i/n - \mathbb{E}[I_n^i/n]\|_3\right) = O\left(n^{-\frac{1}{2}}\right),$$
$$\|h(1 - I_n^i/n) - \mathbb{E}[h(1 - I_n^i/n)]\|_3 = O\left(\|1 - I_n^i/n - E[1 - I_n^i/n]\|_3\right) = O\left(n^{-\frac{1}{2}}\right),$$

where the $O\left(n^{-\frac{1}{2}}\right)$ upper bound holds by the central limit theorem. A version of the central limit theorem that includes convergence of the moments is given in [23, Theorem 4.2]. Details on the calculation are given in lemma A.2.2.

4.3. LIMIT THEOREMS (CONTRACTION METHOD)

Thus, the asymptotic $\sigma_i(n) \sim \sigma\sqrt{n \log n}$ yields part (a).

Similarly, the Lipschitz-continuity of f_0 and f_1 implies by lemma 3.3.8 that

$$\|f_0(I_n^i) - \mathbb{E}[f_0(I_n^i)]\|_3 = O\left(\|I_n^i - \mathbb{E}[I_n^i]\|_3\right) = O\left(\sqrt{n}\right),$$
$$\|f_1(n - I_n^i) - \mathbb{E}[f_1(n - I_n^i)]\|_3 = O\left(\|n - I_n^i - \mathbb{E}[n - I_n^i]\|_3\right) = O\left(\sqrt{n}\right)$$

where the $O\left(\sqrt{n}\right)$ upper bound holds by the central limit theorem [23, Theorem 4.2]. Once again, the asymptotic of $\sigma_i(n)$ yields (b).

Finally, (c) holds by assumption (1.13) on η_n^i. □

In the spirit of the contraction method presented in section 3.3.2, (weak) limits Y^0 and Y^1 of $(Y_n^0)_{n \geq 0}$ and $(Y_n^1)_{n \geq 0}$ should satisfy

$$\begin{aligned} Y^0 &\stackrel{d}{=} \sqrt{p_{00}} Y^0 + \sqrt{p_{01}} Y^1, \\ Y^1 &\stackrel{d}{=} \sqrt{p_{10}} Y^0 + \sqrt{p_{11}} Y^1, \end{aligned} \quad (4.31)$$

with Y^0 and Y^1 independent.

The corresponding limit map T is defined as

$$\begin{aligned} T : \mathfrak{P} \times \mathfrak{P} &\to \mathfrak{P} \times \mathfrak{P}, \\ \begin{pmatrix} \rho_1 \\ \rho_2 \end{pmatrix} &\mapsto \begin{pmatrix} \mathcal{L}(\sqrt{p_{00}} Z_1 + \sqrt{p_{10}} Z_2) \\ \mathcal{L}(\sqrt{p_{10}} Z_1 + \sqrt{p_{11}} Z_2) \end{pmatrix}, \end{aligned} \quad (4.32)$$

where Z_1 and Z_2 are independent and $\mathcal{L}(Z_1) = \rho_1$, $\mathcal{L}(Z_2) = \rho_2$.

Such a limit map was already introduced in section 3.3.2. There, the Zolotarev metric ζ_s was generalized to a metric $\zeta_s^\vee$ on $\mathfrak{P}_s(0,1) \times \mathfrak{P}_s(0,1)$ for $s \in (2,3]$ by taking the maximum of the ζ_s-distances in each component. Recall that T is a contracting map on $\mathfrak{P}_s(0,1) \times \mathfrak{P}_s(0,1)$ with respect to $\zeta_s^\vee$. The unique fixed point of T in $\mathfrak{P}_s(0,1) \times \mathfrak{P}_s(0,1)$ is a pair of standard normal distributions $(\mathcal{N}(0,1), \mathcal{N}(0,1))$.

Also recall that $\zeta_s(X, Y) := \zeta_s(\mathcal{L}(X), \mathcal{L}(Y))$ for random variables X, Y.

Proof of theorem 4.3.1. Convergence in $\zeta_s^\vee$ implies weak convergence in each component (lemma 3.3.2). Therefore, it is sufficient to show that for some $s \in (2,3]$, as $n \to \infty$,

$$\begin{aligned} \zeta_s\left(Y_n^0, N_0\right) &\longrightarrow 0, \\ \zeta_s\left(Y_n^1, N_1\right) &\longrightarrow 0 \end{aligned} \quad (4.33)$$

with $\mathcal{L}(N_i) = \mathcal{N}(0,1)$ for both $i \in \Sigma$.

To this end, fix $s \in (2,3]$ such that $\|Y_n^i\|_s < \infty$ for all $n \in \mathbb{N}_0$ and $i \in \Sigma$ (which exists by condition (1.11)). Let n_0 be chosen in such a way that

$$\sigma_i(n) > 0 \quad \text{for all } n \geq n_0, \, i \in \Sigma.$$

Such an integer exists because $\sigma_i(n) \to \infty$ for both $i \in \Sigma$ by theorem 4.2.1.

In order to handle the convergence of the coefficients in the distributional equations (4.30) separately, one usually introduces a accompanying sequence (cf. the analysis of the *Bernoulli Source Model* in [54]).

68 CHAPTER 4. MOMENTS AND LIMIT THEOREMS

Here, two accompanying sequences $(Q_n^0)_{n\geq 0}$ and $(Q_n^1)_{n\geq 0}$ are defined as

$$Q_{n+d}^i := \frac{\sigma_0(I_n^i)}{\sigma_i(n+d)} N_0 + \frac{\sigma_1(n-I_n^i)}{\sigma_i(n+d)} N_1 + b_i(n), \qquad n \geq n_0 - d, \, i \in \Sigma, \qquad (4.34)$$

where N_0, N_1 follow the standard normal distribution $\mathcal{N}(0,1)$, $(I_n^i, b_i(n))_{i \in \Sigma}$ are defined as in (4.30) and N_0, N_1 and $(I_n^i, b_i(n))_{i \in \Sigma}$ are independent. The definition of Q_n^i for $n < n_0$ does not matter for the asymptotic analysis.

Recall that, for $s \in (2, 3]$,

$$\mathfrak{P}_s(0,1) := \{\mathcal{L}(X) \in \mathfrak{P}_s : \mathbb{E}[X] = 0, \text{Var}(X) = 1\}$$

where $\mathfrak{P}_s$ denotes the set of all probability distributions on $\mathbb{R}$ with finite s-th moment.

Note that the accompanying sequences satisfy $\mathcal{L}(Q_n^0), \mathcal{L}(Q_n^1) \in \mathfrak{P}_s(0,1)$ for $n \geq n_0$:

- since $(Y_n^0)_{n \geq 0}$ and $(Y_n^1)_{n \geq 0}$ are centered, the expectation of (4.30) yields $\mathbb{E}[b_i(n)] = 0$ and therefore $\mathbb{E}[Q_n^i] = 0$,

- comparing the second moments in (4.30) conditioned on I_n^i with the conditioned second moments in (4.34) yields $\text{Var}(Q_n^i) = \text{Var}(Y_n^i) = 1$ because

  - $\mathbb{E}[Y_k^j] = 0 = \mathbb{E}[N_j]$ and $\text{Var}(Y_k^j) = 1 = \text{Var}(N_j)$ if $\sigma_j(k) > 0$,
  - $\sigma_j(k)/\sigma_i(n+d) = 0$ otherwise,

- $\|Q_n^i\|_s \leq \|Q_n^i\|_3 < \infty$ by lemma 4.3.2 and $\mathbb{E}[|N_0|^3] = \mathbb{E}[|N_1|^3] < \infty$.

Hence, the distances $\zeta_s(Y_n^i, Q_n^i), \zeta_s(Q_n^i, N_i)$ and $\zeta_s(Y_n^i, N_i)$ are finite for all $n \geq n_0$ and $i \in \Sigma$ with ζ_s defined in (3.24).

The proof of the assertion is split into two parts that yield (4.33) by the triangle inequality:

(a) $\zeta_s^\vee \left(\begin{pmatrix} \mathcal{L}(Q_{n+d}^0) \\ \mathcal{L}(Q_{n+d}^1) \end{pmatrix}, \begin{pmatrix} \mathcal{L}(N_0) \\ \mathcal{L}(N_1) \end{pmatrix} \right) \xrightarrow{n \to \infty} 0,$

(b) $\zeta_s^\vee \left(\begin{pmatrix} \mathcal{L}(Y_{n+d}^0) \\ \mathcal{L}(Y_{n+d}^1) \end{pmatrix}, \begin{pmatrix} \mathcal{L}(Q_{n+d}^0) \\ \mathcal{L}(Q_{n+d}^1) \end{pmatrix} \right) \xrightarrow{n \to \infty} 0.$

For part (a) note that the convergence of $\|\sigma_0(I_n^i)/\sigma_i(n+d)\|_s$, $\|\sigma_1(n-I_n^i)/\sigma_i(n+d)\|_s$ and $\|b_i(n)\|_s$ imply that $\|Q_{n+d}^i\|_s$ is uniformly bounded in n for both $i \in \Sigma$. Hence, by lemma 3.3.5 it is sufficient to show that

$$\ell_s(Q_{n+d}^i, N_i) \to 0, \quad i \in \Sigma.$$

Since $(\mathcal{L}(N_0), \mathcal{L}(N_1))$ is a fixed point of the limit map T, the *Wasserstein* distance is bounded by the L_s-distance of Q_{n+d}^i and $\sqrt{p_{i0}} N_0 + \sqrt{p_{i1}} N_1$ which yields

$$\ell_s(Q_{n+d}^i, N_i) \leq \left\| \left(\frac{\sigma_0(I_n^i)}{\sigma_i(n+d)} - \sqrt{p_{i0}} \right) N_0 \right\|_s + \left\| \left(\frac{\sigma_1(n-I_n^i)}{\sigma_i(n+d)} - \sqrt{p_{i1}} \right) N_1 \right\|_s + \|b_i(n)\|_s$$

$$= \left\| \frac{\sigma_0(I_n^i)}{\sigma_i(n+d)} - \sqrt{p_{i0}} \right\|_s \|N_0\|_s + \left\| \frac{\sigma_1(n-I_n^i)}{\sigma_i(n+d)} - \sqrt{p_{i1}} \right\|_s \|N_1\|_s + \|b_i(n)\|_s. \quad (4.35)$$

This bound, $\|N_0\|_s = \|N_1\|_s < \infty$ and the asymptotic behavior of the coefficients (lemma 4.3.2) imply $\ell_s(Q_{n+d}^i, N_i) \to 0$, $i \in \Sigma$ which yields (a).

4.3. LIMIT THEOREMS (CONTRACTION METHOD)

For part (b) first note that (4.30) and (4.34) imply for both $i \in \Sigma$ and all $n \geq 2n_0$

$$\zeta_s(Y_{n+d}^i, Q_{n+d}^i)$$
$$= \zeta_s\left(\frac{\sigma_0(I_n^i)}{\sigma_i(n+d)}Y_{I_n^i}^0 + \frac{\sigma_1(n-I_n^i)}{\sigma_i(n+d)}Y_{n-I_n^i}^1 + b_i(n), \frac{\sigma_0(I_n^i)}{\sigma_i(n+d)}N_0 + \frac{\sigma_1(n-I_n^i)}{\sigma_i(n+d)}N_1 + b_i(n)\right)$$
$$\overset{(*)}{\leq} \sum_{k=0}^n \mathbb{P}(I_n^i = k)\zeta_s\left(\frac{\sigma_0(k)}{\sigma_i(n+d)}Y_k^0 + \frac{\sigma_1(n-k)}{\sigma_i(n+d)}Y_{n-k}^1 + \tilde{b}_i(n,k),\right.$$
$$\left.\frac{\sigma_0(k)}{\sigma_i(n+d)}N_0 + \frac{\sigma_1(n-k)}{\sigma_i(n+d)}N_1 + \tilde{b}_i(n,k)\right)$$

where $\tilde{b}_i(n,k)$ is independent of $(Y_k^0, Y_{n-k}^1, N_0, N_1)$ and $\mathcal{L}(\tilde{b}_i(n,k)) = \mathcal{L}(b_i(n)|I_n^i = k)$ for every $k \in \{0, \ldots n\}$. More precisely, the assumption on the *toll term* yields that $\tilde{b}_i(n,k)$ is a constant that occurs when I_n^i is replaced by k in the definition of $b_i(n)$.

The upper bound (*) holds by conditioning on I_n^i in the definition of ζ_s and then using Jensen's inequality.

Note that either $\sigma_0(k) > 0$ or $\sigma_1(n-k) > 0$ for $n \geq 2n_0$. Hence, splitting the sum into $J_1 := \{k \in \{0, \ldots, n\} : \sigma_0(k) = 0\}$, $J_2 := \{k \in \{0, \ldots, n\} : \sigma_1(n-k) = 0\}$ and $J_3 = \{0, \ldots, n\} \setminus (J_1 \cup J_2)$ yields

$$\zeta_s(Y_{n+d}^i, Q_{n+d}^i)$$
$$\leq \sum_{k \in J_1} \mathbb{P}(I_n^i = k)\zeta_s\left(\frac{\sigma_1(n-k)}{\sigma_i(n+d)}Y_{n-k}^1 + \tilde{b}_i(n), \frac{\sigma_1(n-k)}{\sigma_i(n+d)}N_1 + \tilde{b}_i(n)\right)$$
$$+ \sum_{k \in J_2} \mathbb{P}(I_n^i = k)\zeta_s\left(\frac{\sigma_0(k)}{\sigma_i(n+d)}Y_k^0 + \tilde{b}_i(n), \frac{\sigma_0(k)}{\sigma_i(n+d)}N_0 + \tilde{b}_i(n)\right)$$
$$+ \sum_{k \in J_3} \mathbb{P}(I_n^i = k)\zeta_s\left(\frac{\sigma_0(k)}{\sigma_i(n+d)}Y_k^0 + \frac{\sigma_1(n-k)}{\sigma_i(n+d)}Y_{n-k}^1 + \tilde{b}_i(n),\right.$$
$$\left.\frac{\sigma_0(k)}{\sigma_i(n+d)}N_0 + \frac{\sigma_1(n-k)}{\sigma_i(n+d)}N_1 + \tilde{b}_i(n)\right).$$

With the notation

$$d_i(n) := \begin{cases} \zeta_s(Y_n^i, N_i), & \text{if } \sigma_i(n) > 0, \\ 0, & \text{otherwise,} \end{cases}$$

one obtains by lemma 3.3.3 and corollary 3.3.4 that

$$\zeta_s(Y_{n+d}^i, Q_{n+d}^i) \leq \mathbb{E}\left[\left(\frac{\sigma_1(n-I_n^i)}{\sigma_i(n+d)}\right)^s d_1(n-I_n^i)\mathbb{1}_{\{I_n^i \in J_1\}}\right] + \mathbb{E}\left[\left(\frac{\sigma_0(I_n^i)}{\sigma_i(n+d)}\right)^s d_0(I_n^i)\mathbb{1}_{\{I_n^i \in J_2\}}\right]$$
$$+ \mathbb{E}\left[\left(\left(\frac{\sigma_0(I_n^i)}{\sigma_i(n+d)}\right)^s d_0(I_n^i) + \left(\frac{\sigma_1(n-I_n^i)}{\sigma_i(n+d)}\right)^s d_1(n-I_n^i)\right)\mathbb{1}_{\{I_n^i \in J_3\}}\right]$$
$$= \mathbb{E}\left[\left(\frac{\sigma_0(I_n^i)}{\sigma_i(n+d)}\right)^s d_0(I_n^i) + \left(\frac{\sigma_1(n-I_n^i)}{\sigma_i(n+d)}\right)^s d_1(n-I_n^i)\right]. \quad (4.36)$$

Now let $d(n) := \max\{d_0(n), d_1(n)\}$. Note that, for all $n \geq n_0$,

$$d(n) = \zeta_s^\vee\left(\begin{pmatrix}\mathcal{L}(Y_n^0)\\ \mathcal{L}(Y_n^1)\end{pmatrix}, \begin{pmatrix}\mathcal{L}(N_0)\\ \mathcal{L}(N_1)\end{pmatrix}\right)$$

and therefore, it is sufficient to show that $d(n) \to 0$. The triangle inequality, part (a) and taking the maximum in (4.36) yield for both $i \in \Sigma$ and all $n \geq 2n_0$

$$d_i(n+d) \leq \mathbb{E}\left[\left(\left(\frac{\sigma_0(I_n^i)}{\sigma_i(n+d)}\right)^s + \left(\frac{\sigma_1(n-I_n^i)}{\sigma_i(n+d)}\right)^s\right)\mathbb{1}_{\{I_n^i \in \{1,\ldots,n-1\}\}}\right]\sup_{0\leq j\leq n-1} d(j)$$
$$+ \left(p_{i0}^n\left(\frac{\sigma_0(n)}{\sigma_i(n+d)}\right)^s + p_{i1}^n\left(\frac{\sigma_1(n)}{\sigma_i(n+d)}\right)^s\right)d(n) + o(1).$$

With the notation

$$\xi(n) := \max_{i\in\Sigma}\mathbb{E}\left[\left(\left(\frac{\sigma_0(I_n^i)}{\sigma_i(n+d)}\right)^s + \left(\frac{\sigma_1(n-I_n^i)}{\sigma_i(n+d)}\right)^s\right)\mathbb{1}_{\{I_n^i \in \{1,\ldots,n-1\}\}}\right],$$
$$\varepsilon(n) := \max_{i\in\Sigma}\left(p_{i0}^n\left(\frac{\sigma_0(n)}{\sigma_i(n+d)}\right)^s + p_{i1}^n\left(\frac{\sigma_1(n)}{\sigma_i(n+d)}\right)^s\right),$$

one obtains

$$d(n+d) \leq \xi(n)\sup_{0\leq j\leq n-1} d(j) + \varepsilon(n)d(n) + o(1).$$

Note that the asymptotic behavior of the coefficients (lemma 4.3.2) and $\sigma_i(n) \sim \sigma\sqrt{n\log n}$ (theorem 4.2.1) imply for $\xi(n)$ and $\varepsilon(n)$ the following asymptotic as $n \to \infty$:

$$\xi(n) \to \xi := \max_{i\in\Sigma}\left(p_{i0}^{s/2} + p_{i1}^{s/2}\right), \qquad \varepsilon(n) \to 0. \qquad (4.37)$$

In particular, one obtains $\varepsilon(n) < 1$ for large n and therefore,

$$d(n+d) \leq \begin{cases} \frac{\xi(n)}{1-\varepsilon(n)}\sup_{0\leq j\leq n-1} d(j) + o(1), & \text{if } d = 0, \\ (\xi(n) + \varepsilon(n))\sup_{0\leq j\leq n} d(j) + o(1), & \text{if } d \geq 1. \end{cases} \qquad (4.38)$$

This bound, $\xi(n)/(1-\varepsilon(n)) \to \xi < 1$ and $\xi(n) + \varepsilon(n) \to \xi < 1$ imply by induction on n that $(d(n))_{n\geq 0}$ is bounded.

Now let $\eta := \sup_{n\in\mathbb{N}_0} d(n)$ and $\lambda := \limsup_{n\to\infty} d(n)$. Moreover, for any $\varepsilon > 0$, let $m_\varepsilon \geq n_0$ be chosen in such a way that $d(n) \leq \lambda + \varepsilon$ for all $n \geq m_\varepsilon$. Then, (4.36) (and part (a) combined with the triangle inequality) implies for $n > 2m_\varepsilon$ that

$$d_i(n+d) \leq \mathbb{E}\left[\left(\left(\frac{\sigma_0(I_n^i)}{\sigma_i(n+d)}\right)^s + \left(\frac{\sigma_1(n-I_n^i)}{\sigma_i(n+d)}\right)^s\right)\mathbb{1}_{\{I_n^i \notin \{m_\varepsilon,\ldots,n-m_\varepsilon\}\}}\right]\eta + \xi(n)(\lambda+\varepsilon) + o(1)$$
$$= \xi(n)(\lambda+\varepsilon) + o(1).$$

Maximizing over $i \in \Sigma$ and letting $n \to \infty$ yields

$$\lambda \leq \xi(\lambda + \varepsilon).$$

Recall that $\xi < 1$ and that $\varepsilon > 0$ may be chosen arbitrarily small. Hence, the bound implies that $\lambda = 0$. Therefore,

$$\lim_{n\to\infty} \zeta_s^\vee\left(\begin{pmatrix}Y_{n+d}^0\\Y_{n+d}^1\end{pmatrix},\begin{pmatrix}N_0\\N_1\end{pmatrix}\right) = \lim_{n\to\infty} d(n) = 0$$

which implies weak convergence of each component by lemma 3.3.2. $\square$

4.4 Transfer to Arbitrary Initial Distributions

Recall that the results so far only concerned the sequences $(X_n^0)_{n\geq 0}$ and $(X_n^1)_{n\geq 0}$ that correspond to the initial distributions $p_{i0}\delta_0 + p_{i1}\delta_1$, $i \in \Sigma$.
A transfer to $(X_n^\mu)_{n\geq 0}$ with arbitrary initial distribution $\mu = \mu_0\delta_0 + \mu_1\delta_1$, $\mu_0 \in [0,1]$, $\mu_1 = 1-\mu_0$, relies on the distributional equation (1.7) which is

$$X_{n+d}^\mu \stackrel{d}{=} X_{K_n^\mu}^0 + X_{n-K_n^\mu}^1 + \eta_n^\mu, \qquad n \in \mathbb{N} \tag{4.39}$$

where $(X_n^0)_{n\geq 0}$, $(X_n^1)_{n\geq 0}$ and (K_n^μ, η_n^μ) are independent and K_n^μ follows the binomial distribution $B(n, \mu(0))$.

Recall that the assumptions (1.12) on the toll term η_n^μ are

$$\mathbb{E}[\eta_n^\mu] = O(n), \qquad \mathrm{Var}(\eta_n^\mu) = O(n) \quad (n \to \infty)$$

whereas in the special cases $\mu = p_{i0}\delta_0 + p_{i1}\delta_1$, $i \in \Sigma$ the toll terms satisfy the additional conditions (1.13).

With the conditions above, the following holds for any initial distribution μ and any transition matrix P that satisfies (1.9):

Theorem 4.4.1. *Mean and variance of X_n^μ satisfy, as $n \to \infty$,*

$$\mathbb{E}[X_n^\mu] = \frac{1}{H} n \log n + O(n), \qquad \mathrm{Var}(X_n^\mu) \sim \sigma^2 n \log n$$

where H is the entropy rate of the source defined in (1.4) and $\sigma^2 > 0$ is given by

$$\sigma^2 = \frac{\pi_0 p_{00} p_{01}}{H^3}\left(\log(p_{00}/p_{01}) + \frac{H_1 - H_0}{p_{01} + p_{10}}\right)^2 + \frac{\pi_1 p_{10} p_{11}}{H^3}\left(\log(p_{10}/p_{11}) + \frac{H_1 - H_0}{p_{01} + p_{10}}\right)^2.$$

Moreover, as $n \to \infty$,

$$\frac{X_n^\mu - \mathbb{E}[X_n^\mu]}{\sqrt{\mathrm{Var}(X_n^\mu)}} \xrightarrow{d} \mathcal{N}(0,1)$$

where $\mathcal{N}(0,1)$ denotes a random variable with the standard normal distribution.

Proof. For $i \in \Sigma$ and $n \in \mathbb{N}_0$ let

$$\nu_i(n) := \mathbb{E}[X_n^i], \qquad V_i(n) := \mathrm{Var}(X_n^i).$$

Note that, similarly to lemma 3.0.4, the distributional equation (4.39) and the independence therein implies

$$\mathbb{E}[X_{n+d}^\mu] = \mathbb{E}[\nu_0(K_n^\mu) + \nu_1(n - K_n^\mu)] + \mathbb{E}[\eta_n^\mu]$$

with $\mathbb{E}[\eta_n^\mu] = O(n)$ by condition (1.12).

Theorem 4.1.1 yields $\nu_i(n) = \frac{1}{H} n \log n + O(n)$ and therefore,

$$\mathbb{E}[X_{n+d}^\mu] = \frac{1}{H}\mathbb{E}[K_n^\mu \log(K_n^\mu) + (n - K_n^\mu)\log(n - K_n^\mu)] + O(n)$$

$$= \frac{1}{H} n \log n + \frac{n}{H}\mathbb{E}\left[\frac{K_n^\mu}{n}\log\left(\frac{K_n^\mu}{n}\right) + \left(1 - \frac{K_n^\mu}{n}\right)\log\left(1 - \frac{K_n^\mu}{n}\right)\right] + O(n).$$

Note that $x \mapsto x \log x + (1-x) \log(1-x)$ is bounded on $[0,1]$ (with the convention $0 \log 0 := 0$). Hence, the equation above yields

$$\mathbb{E}[X_{n+d}^\mu] = \frac{1}{H} n \log n + \mathrm{O}(n)$$

and the assertion follows since $(n+d) \log(n+d) = n \log n + \mathrm{O}(\log n)$.

For the variance note that by the same arguments as in lemma 3.0.4

$$\begin{aligned}\mathrm{Var}(X_{n+d}^\mu) &= \mathbb{E}[V_0(K_n^\mu)] + \mathbb{E}[V_1(n - K_n^\mu)] + \mathrm{Var}(\nu_0(K_n^\mu) + \nu_1(n - K_n^\mu) + \eta_n^\mu) \\ &= \sigma^2 \mathbb{E}[K_n^\mu \log(K_n^\mu) + (n - K_n^\mu) \log(n - K_n^\mu)] \\ &\quad + \mathrm{Var}(\nu_0(K_n^\mu) + \nu_1(n - K_n^\mu) + \eta_n^\mu) + \mathrm{O}\left(n \sqrt{\log n}\right)\end{aligned}$$

where the last equality holds by theorem 4.2.1. Hence, it is sufficient to show

(a) $\mathbb{E}[K_n^\mu \log(K_n^\mu) + (n - K_n^\mu) \log(n - K_n^\mu)] = n \log n + \mathrm{O}(n)$,

(b) $\mathrm{Var}(\nu_0(K_n^\mu) + \nu_1(n - K_n^\mu) + \eta_n^\mu) = \mathrm{O}(n)$.

Part (a) was already shown in the analysis of $\mathbb{E}[X_{n+d}^\mu]$. For part (b) note that theorem 4.1.1 yields $\nu_i(n) = \frac{1}{H} n \log n + f_i(n)$, $i \in \Sigma$, where f_0 and f_1 are Lipschitz-continuous. Therefore, part (b) follows by lemma 4.2.2 and the bounds

(b1) $\mathrm{Var}(K_n^\mu \log(K_n^\mu) + (n - K_n^\mu) \log(n - K_n^\mu)) = \mathrm{O}(n)$,

(b2) $\mathrm{Var}(f_0(K_n^\mu)) = \mathrm{O}(n)$, $\mathrm{Var}(f_1(n - K_n^\mu)) = \mathrm{O}(n)$,

(b3) $\mathrm{Var}(\eta_n^\mu) = \mathrm{O}(n)$.

For (b1) note that

$$\begin{aligned}&\sqrt{\mathrm{Var}(K_n^\mu \log(K_n^\mu) + (n - K_n^\mu) \log(n - K_n^\mu))} \\ &= \|K_n^\mu \log(K_n^\mu) - \mathbb{E}[K_n^\mu \log(K_n^\mu)] + (n - K_n^\mu) \log(n - K_n^\mu) - \mathbb{E}[(n - K_n^\mu) \log(n - K_n^\mu)]\|_2 \\ &= \|K_n^\mu \log(K_n^\mu/n) - \mathbb{E}[K_n^\mu \log(K_n^\mu/n)] + (n - K_n^\mu) \log(1 - K_n^\mu/n) - \mathbb{E}[(n - K_n^\mu) \log(1 - K_n^\mu/n)]\|_2 \\ &\leq n \left(\|h(K_n^\mu/n) - \mathbb{E}[h(K_n^\mu/n)]\|_2 + \|h(1 - K_n^\mu/n) - \mathbb{E}[h(1 - K_n^\mu/n)]\|_2\right)\end{aligned}$$

where $h(x) := x \log x$ (and $h(0) = 0$).

Bounds on $\|h(K_n^\mu/n) - \mathbb{E}[h(K_n^\mu/n)]\|_2$ and $\|h(1 - K_n^\mu/n) - \mathbb{E}[h(1 - K_n^\mu/n)]\|_2$ may be computed as in the analysis of the coefficients in lemma 4.3.2. More precisely, an easy calculation given in lemma A.2.2 reveals that

$$\|h(K_n^\mu/n) - \mathbb{E}[h(K_n^\mu/n)]\|_2 = \mathrm{O}\left(n^{-\frac{1}{2}}\right), \qquad \|h(1 - K_n^\mu/n) - \mathbb{E}[h(1 - K_n^\mu/n)]\|_2 = \mathrm{O}\left(n^{-\frac{1}{2}}\right)$$

and therefore, (b1) holds.

Moreover, (b2) follows by the Lipschitz-continuity of f_0 and f_1 and lemma 3.3.8.

Finally, (b3) holds by assumption (1.12) on η_n^μ which finishes the asymptotic analysis of the variance.

It remains to show that, as $n \to \infty$,

$$\frac{X_n^\mu - \mathbb{E}[X_n^\mu]}{\sqrt{\mathrm{Var}(X_n^\mu)}} \xrightarrow{d} \mathcal{N}(0,1).$$

4.4. TRANSFER TO ARBITRARY INITIAL DISTRIBUTIONS

To this end, consider the normalized random variables $(Y_n^0)_{n\geq 0}$ and $(Y_n^1)_{n\geq 0}$ defined as

$$Y_n^i := \begin{cases} \frac{X_n^i - \mathbb{E}[X_n^i]}{\sqrt{\operatorname{Var}(X_n^i)}}, & \text{if } \operatorname{Var}(X_n^i) > 0, \\ 0, & \text{otherwise.} \end{cases}$$

Theorem 4.3.1 yields $Y_n^i \xrightarrow{d} \mathcal{N}(0,1)$ for both $i \in \Sigma$. Moreover, the distributional equation (4.39) implies that

$$\frac{X_{n+d}^\mu - \mathbb{E}[X_{n+d}^\mu]}{\sqrt{\operatorname{Var}(X_{n+d}^\mu)}} \stackrel{d}{=} \sqrt{\frac{V_0(K_n^\mu)}{\operatorname{Var}(X_{n+d}^\mu)}} Y_{K_n^\mu}^0 + \sqrt{\frac{V_1(n - K_n^\mu)}{\operatorname{Var}(X_{n+d}^\mu)}} Y_{n-K_n^\mu}^1 + b_\mu(n)$$

where $b_\mu(n)$ is given by

$$b_\mu(n) = \frac{\nu_0(K_n^\mu) - \mathbb{E}[\nu_0(K_n^\mu)] + \nu_1(n - K_n^\mu) - \mathbb{E}[\nu_1(n - K_n^\mu)] + \eta_n^\mu - \mathbb{E}[\eta_n^\mu]}{\sqrt{\operatorname{Var}(X_{n+d}^\mu)}}.$$

By Slutsky's theorem it is sufficient to show that, as $n \to \infty$,

(I) $\sqrt{\frac{V_0(K_n^\mu)}{\operatorname{Var}(X_{n+d}^\mu)}} Y_{K_n^\mu}^0 + \sqrt{\frac{V_1(n - K_n^\mu)}{\operatorname{Var}(X_{n+d}^\mu)}} Y_{n-K_n^\mu}^1 \xrightarrow{d} \mathcal{N}(0,1)$,

(II) $b_\mu(n) \xrightarrow{P} 0$.

Part (I) is a consequence of the independence and asymptotic normality of $(Y_n^0)_{n\geq 0}$ and $(Y_n^1)_{n\geq 0}$: For $\mu_0 = 1$ (or $\mu_0 = 0$) part (I) directly follows from the asymptotic normality of Y_n^0 (or Y_n^1) and the asymptotic behavior of the variances.

Now let $\mu_0 \in (0,1)$ and $A_n := [\mu_0 n - n^{2/3}, \mu_0 n + n^{2/3}] \cap \mathbb{N}_0$. Chernoff's bound on the binomial distribution (or the central limit theorem) implies that $\mathbb{P}(K_n^\mu \in A_n) \to 1$ and therefore, for any $x \in \mathbb{R}$,

$$\mathbb{P}\left(\sqrt{\frac{V_0(K_n^\mu)}{\operatorname{Var}(X_{n+d}^\mu)}} Y_{K_n^\mu}^0 + \sqrt{\frac{V_1(n - K_n^\mu)}{\operatorname{Var}(X_{n+d}^\mu)}} Y_{n-K_n^\mu}^1 \leq x\right)$$

$$= o(1) + \sum_{j \in A_n} \mathbb{P}(K_n^\mu = j) \mathbb{P}\left(\sqrt{\frac{V_0(j)}{\operatorname{Var}(X_{n+d}^\mu)}} Y_j^0 + \sqrt{\frac{V_1(n - j)}{\operatorname{Var}(X_{n+d}^\mu)}} Y_{n-j}^1 \leq x\right)$$

Also note that, for $j \in A_n$,

$$\sqrt{V_0(j)/\operatorname{Var}(X_{n+d}^\mu)} \to \sqrt{\mu(0)}, \quad \sqrt{V_1(n-j)/\operatorname{Var}(X_{n+d}^\mu)} \to \sqrt{1 - \mu(0)}$$

and $(Y_j^0, Y_{n-j}^1) \xrightarrow{d} (N_0, N_1)$ where N_0 and N_1 are two independent $\mathcal{N}(0,1)$ distributed random variables. The convolution property $\sqrt{\mu(0)} N_0 + \sqrt{1 - \mu(0)} N_1 \stackrel{d}{=} N_0$ yields

$$\mathbb{P}\left(\sqrt{\frac{V_0(K_n^\mu)}{\operatorname{Var}(X_{n+d}^\mu)}} Y_{K_n^\mu}^0 + \sqrt{\frac{V_1(n - K_n^\mu)}{\operatorname{Var}(X_{n+d}^\mu)}} Y_{n-K_n^\mu}^1 \leq x\right)$$

$$= o(1) + \sum_{j \in A_n} \mathbb{P}(K_n^\mu = j)(\mathbb{P}(N_0 \leq x) + o(1))$$

$$= \mathbb{P}(N_0 \leq x) + o(1)$$

where the last equality is justified by the dominated convergence theorem. Since this holds for any $x \in \mathbb{R}$, part (I) follows.

Part (II) is a consequence of Markov's inequality and the following result which is shown next:

$$\|b_\mu(n)\|_1 \to 0 \quad (n \to \infty). \tag{4.40}$$

Recall that $\nu_i(n) = n\log n/H + f_i(n)$, $i \in \Sigma$, by theorem 4.1.1 with Lipschitz-continuous functions f_0 and f_1. Hence, with the notation $h(x) := x\log x$ for $x > 0$ and $h(0) = 0$,

$$\begin{aligned}
\|b_\mu(n)\|_1 &= \left\|\frac{\nu_0(K_n^\mu) - \mathbb{E}[\nu_0(K_n^\mu)] + \nu_1(n - K_n^\mu) - \mathbb{E}[\nu_1(n - K_n^\mu)] + \eta_n^\mu - \mathbb{E}[\eta_m^\mu]}{\sqrt{\mathrm{Var}(X_{n+d}^\mu)}}\right\|_1 \\
&\leq \frac{1}{H\sqrt{\mathrm{Var}(X_{n+d}^\mu)}}(\|h(K_n^\mu) - \mathbb{E}[h(K_n^\mu)] + h(n - K_n^\mu) - \mathbb{E}[h(n - K_n^\mu)]\|_1) \\
&\quad + \frac{1}{\sqrt{\mathrm{Var}(X_{n+d}^\mu)}}(\|f_0(K_n^\mu) - \mathbb{E}[f_0(K_n^\mu)] + f_1(n - K_n^\mu) - \mathbb{E}[f_1(n - K_n^\mu)]\|_1) \\
&\quad + \frac{1}{\sqrt{\mathrm{Var}(X_{n+d}^\mu)}}\|\eta_n^\mu - \mathbb{E}[\eta_n^\mu]\|_1.
\end{aligned}$$

Therefore, it only remains to show that, as $n \to \infty$

(II.1) $\|h(K_n^\mu) - \mathbb{E}[g(K_n^\mu)] + h(n - K_n^\mu) - \mathbb{E}[g(n - K_n^\mu)]\|_1 = o(\sqrt{n\log n})$,

(II.2) $\|f_0(K_n^\mu) - \mathbb{E}[f_0(K_n^\mu)]\|_1 = o(\sqrt{n\log n})$, $\quad \|f_1(n - K_n^\mu) - \mathbb{E}[f_1(n - K_n^\mu)]\|_1 = o(\sqrt{n\log n})$,

(II.3) $\|\eta_n^\mu - \mathbb{E}[\eta_n^\mu]\|_1 = o(\sqrt{n\log n})$.

The bounds (II.1)-(II.3) follow from (b1)-(b3) in the asymptotic analysis of the variance and the fact that, by Jensen's inequality,

$$\|X - \mathbb{E}[X]\|_1 \leq \|X - \mathbb{E}[X]\|_2 = \sqrt{\mathrm{Var}(X)}$$

for any real valued random variable X. Finally, combining (II.1)-(II.3) and Markov's inequality yields $b_\mu(n) \xrightarrow{\mathbb{P}} 0$ and therefore the asymptotic normality by part (I) and Slutsky's theorem. $\square$

Chapter 5

The Radix Selection Algorithm

The algorithm *Radix Select* is a one-sided version of the sorting algorithm introduced in section 1.1. Given a list $\mathcal{X} = [\Xi_1, \ldots, \Xi_n]$ of strings (or binary expansions of numbers) and a rank $k \in \{1, \ldots, n\}$, *Radix Select* returns the element $\Xi_{(k)}$ of rank k. Here, the ranking of strings is determined by their lexicographical order.

There are several stochastic models for $\mathcal{X}$ summarized in section 1.2. The results in this chapter focus on the *Markov Source Model*. Recall that strings in this model are considered to be independent and identically distributed. The distribution of each string is given by a Markov chain with an arbitrary initial distribution μ and a transition matrix $P = (p_{ij})_{i,j \in \{0,1\}}$.

In addition to the model for the list $\mathcal{X}$, the analysis of *Radix Select* requires a model for the rank k. In a rather simple model, the rank is assumed to be independent of the list $\mathcal{X}$ and uniformly distributed on $\{1, \ldots, n\}$. Such a model was proposed and studied for a list generated by a *symmetric Bernoulli Source* in [15]. Due to the fact that such a model averages the complexity over the possible ranks, the model is called the *Grand Averages Model*.

A more detailed study of the complexity is given when considering all ranks $\lfloor tn \rfloor + 1$ simultaneously as a process in $t \in [0, 1)$. Results on that process may be transferred to the *Grand Averages Model* by choosing $t = U$ where U is uniformly distributed on $[0, 1)$ and independent of the input list $\mathcal{X}$. For further references, a model that considers all ranks $\lfloor tn \rfloor + 1$, $t \in [0, 1)$, is called the *Quantile Model*.

The next section gives an introduction to the *Radix Select* algorithm. Afterwards, the average complexity in the *Quantile Model* is studied for *Markov Sources*. The study of the *Quantile Model* also requires a bound on the average worst case complexity of *Radix Select* which is done in section 5.2.1. Finally, an asymptotic expansion of the mean and a weak limit for the complexity in the *Grand Averages Model* with *Markov Sources* are given in section 5.3. Section 5.3.3 provides an explanation why the complexity in the *Grand Averages Model* is less concentrated when sources other than the *symmetric Bernoulli Source* are considered.

The results in this chapter were recently presented in the Analysis of Algorithms conference [47]. Aside from the study of *Markov Sources*, [47] includes a limit law for the complexity in the *Quantile Model* with a *symmetric Bernoulli Source* and several path properties of the limit process.

5.1 Introduction

Let $\mathcal{X} = [\Xi_1, \ldots, \Xi_n]$ be a list of n strings where each string Ξ_i is a sequence $(\xi_k^{(i)})_{k \geq 1}$ of symbols drawn from a finite, ordered alphabet Σ. For simplicity, the symbols in Σ are considered to be renamed such that $\Sigma = \{0, \ldots, b-1\}$ with $b = |\Sigma|$. The elements of $\mathcal{X}$ may be interpreted as words from a given language or b-ary expansions of numbers in the unit interval.

Moreover, let $k \in \{1, \ldots, n\}$ be the rank of the element sought after. Here, the rank of an element is determined by the lexicographical order of $\Xi_1, \ldots, \Xi_n$.

Radix Select is a one-sided version of *Radix Sort* introduced in section 1.1. As in the sorting algorithm, the list $\mathcal{X}$ is split into sublists (*Buckets*) $\mathcal{X}_0, \ldots, \mathcal{X}_{b-1}$ such that Ξ_i is placed in list $\mathcal{X}_j$ if and only if $\xi_1^{(i)} = j$. Afterwards, the algorithm determines the sublist that contains the element of rank k by considering the sizes of the sublists. More precisely, let

$$m(k) := \min\left\{ i \in \{0, \ldots, b-1\} : \sum_{\ell=0}^{i} |\mathcal{X}_\ell| \geq k \right\}$$

where $|\mathcal{X}_\ell|$ denotes the number of strings in sublist $|\mathcal{X}_\ell|$. Then, as long as $\mathcal{X}_{m(k)}$ contains more than one element, the algorithm is recursively applied to the sublist $\mathcal{X}_{m(k)}$ searching for the string with rank $k - \sum_{\ell=0}^{m(k)-1} |\mathcal{X}_\ell|$. The recursive call of the algorithm ignores the first symbol of each element in $\mathcal{X}_{m(k)}$ in the sense that the strings are distributed into sublists according to their second symbol.

Figure 5.1 gives an example for *Radix Select* on six strings. The complexity of the algorithm is measured by the number of *Bucket Operations*. Here, a *Bucket Operation* denotes the placement of a string into a sublist.

For simplicity, all results on *Radix Select* are only derived for the binary alphabet $\Sigma = \{0, 1\}$.

Figure 5.1 The *Radix Select* algorithm on 6 strings searching for rank 2. Only the list that contains the element with rank 2 (green) is recursively split. The total number of *Bucket Operations* is $6 + 3 + 2 \cdot 2 = 13$.

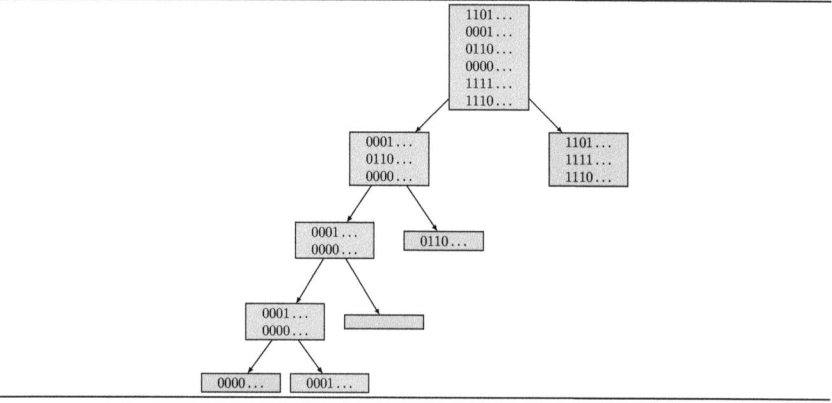

5.1. INTRODUCTION

5.1.1 The Wasserstein Metric

Most of the results are derived by the *Contraction Method* introduced in section 3.3. However, the proofs are done in the *Wasserstein* metric and do not require the *Zolotarev* metric which appeared in the analysis of *Radix Sort*. This section contains all definitions and basic properties needed to derive limit laws in the *Wasserstein* metrics.

Let $p \geq 1$ and X and Y be two real valued random variables with $\mathbb{E}[|X|^p] < \infty$ and $\mathbb{E}[|Y|^p] < \infty$. Then, the *Wasserstein* distance of X and Y is defined as

$$\ell_p(X,Y) := \ell_p(\mathcal{L}(X), \mathcal{L}(Y)) := \inf\{\|W - Z\|_p \;:\; \mathcal{L}(W) = \mathcal{L}(X), \mathcal{L}(Z) = \mathcal{L}(Y)\} \quad (5.1)$$

where the infimum is taken over all random vectors (W, Z) on a common probability space with marginals $\mathcal{L}(W) = \mathcal{L}(X)$ and $\mathcal{L}(Z) = \mathcal{L}(Y)$. Here, $\|\cdot\|_p$ denotes the L_p-norm which, for $p \geq 1$, is given by

$$\|W - Z\|_p = \mathbb{E}[|W - Z|^p]^{1/p}.$$

Recall that $\mathfrak{P}_p$ denotes the set of all probability distributions on $\mathbb{R}$ with finite p-th moment.

Lemma 5.1.1. *Let $p \geq 1$ and $\mathcal{M} \subset \mathfrak{P}_p$ be a family of distributions on $\mathbb{R}$ with finite p-th moment. Then, there exists a family $\{X_\rho \;:\; \rho \in \mathcal{M}\}$ of random variables on a common probability space such that $\mathcal{L}(X_\rho) = \rho$ for all $\rho \in \mathcal{M}$ and*

$$\ell_p(\rho_1, \rho_2) = \mathbb{E}[|X_{\rho_1} - X_{\rho_2}|^p]^{1/p}, \quad \rho_1, \rho_2 \in \mathcal{M}.$$

The family $\{X_\rho \;:\; \rho \in \mathcal{M}\}$ is called an optimal ℓ_p-coupling of $\mathcal{M}$.

Proof. Let $F_\rho : \mathbb{R} \to [0,1]$ denote the distribution function of $\rho \in \mathcal{M}$ which is given by

$$F_\rho(x) = \rho((-\infty, x)), \quad x \in \mathbb{R}.$$

Moreover, let $F_\rho^{-1} : [0,1] \to \mathbb{R}$ denote its generalized inverse given by

$$F_\rho^{-1}(y) = \sup\{x \in \mathbb{R} \;:\; F_\rho(x) \leq y\}, \quad y \in [0,1].$$

Then, $\{F_\rho^{-1}(U) \;:\; \rho \in \mathcal{M}\}$ is an optimal ℓ_p-coupling for $\mathcal{M}$ where U is uniformly distributed on $[0,1]$. More precisely, the following holds for all $\rho_1, \rho_2 \in \mathcal{M}$ by [52, Theorem 8.1]:

$$\ell_p(\rho_1, \rho_2) = \left(\int_0^1 |F_{\rho_1}^{-1}(x) - F_{\rho_2}^{-1}(x)|^p dx\right)^{1/p}.$$

Thus, $\{F_\rho^{-1}(U) \;:\; \rho \in \mathcal{M}\}$ is an optimal ℓ_p coupling since the right hand side coincides with the L_p-distance of $F_{\rho_1}^{-1}(U)$ and $F_{\rho_2}^{-1}(U)$. □

Lemma 5.1.2. *The metric space $(\mathfrak{P}_p, \ell_p)$ is complete for any $p \geq 1$.*

Proof. The completeness follows from the existence of optimal ℓ_p-couplings on a common probability space $(\Omega, \mathfrak{A}, \mathbb{P})$ and the completeness of the space $(L_p(\Omega, \mathfrak{A}, \mathbb{P}), \|\cdot\|_p)$ given by the Riesz-Fischer theorem. □

Corollary 5.1.3. *For $p \geq 1$, let $\ell_p^\vee$ be the the metric on $\mathfrak{P}_p \times \mathfrak{P}_p$ given by*

$$\ell_p^\vee\big((\mu_1, \rho_1), (\mu_2, \rho_2)\big) := \max\{\ell_p(\mu_1, \mu_2), \ell_p(\rho_1, \rho_2)\}, \qquad \mu_1, \mu_2, \rho_1, \rho_2 \in \mathfrak{P}_p.$$

Then, $(\mathfrak{P}_p \times \mathfrak{P}_p, \ell_p^\vee)$ is a complete metric space.

Proof. Let $((\mu_n, \rho_n))_{n \geq 1}$ be a Cauchy sequence in $(\mathfrak{P}_p \times \mathfrak{P}_p, \ell_p^\vee)$. Then, both $(\mu_n)_{n \geq 1}$ and $(\rho_n)_{n \geq 1}$ are Cauchy sequences in $(\mathfrak{P}_p, \ell_p)$ which yields the existence of ℓ_p-limits μ and ρ by lemma 5.1.2. The pair $(\mu, \rho) \in \mathfrak{P}_p \times \mathfrak{P}_p$ is the $\ell_p^\vee$ limit of $((\mu_n, \rho_n))_{n \geq 1}$. □

Lemma 5.1.4. *Let $(\rho_n)_{n \geq 1}$ be a sequence in $\mathfrak{P}_p$ for $p \geq 1$. Assume that there exists a limit $\rho \in \mathfrak{P}_p$ such that*

$$\ell_p(\rho_n, \rho) \longrightarrow 0 \qquad (n \to \infty).$$

Then, as $n \to \infty$,

$$\rho_n \xrightarrow{w} \rho, \qquad \int |x|^p \mathrm{d}\rho_n(x) \longrightarrow \int |x|^p \mathrm{d}\rho(x)$$

where $\xrightarrow{w}$ denotes weak convergence.

Proof. Let $\{X, X_n : n \geq 1\}$ be an optimal ℓ_p coupling of $\{\rho, \rho_n : n \geq 1\}$. Such a coupling exists by lemma 5.1.1. Then, $\|X_n - X\|_p \to 0$ implies convergence in probability by Markov's inequality and, in particular, weak convergence.

Moreover, the triangle inequality implies

$$\big|\|X_n\|_p - \|X\|_p\big| \leq \|X_n - X\|_p \longrightarrow 0 \qquad (n \to \infty)$$

which yields the second part of the assertion. □

5.2 The Quantile Model

In this section, the first order asymptotic of the mean is derived for the complexity of *Radix Select* in the *Quantile Model* with a *Markov Source*. Unlike *Bernoulli Sources*, the asymptotic of the mean is not continuous in t for *Markov Sources* (with $p_{00} \neq p_{10}$) where $\lfloor tn \rfloor + 1$ is the rank sought after. This makes an analysis concerning weak limits more involved than the analysis presented in [47]. In fact, the rescaling known from the analysis of *Bernoulli Sources* does not lead to weak convergence for *Markov Sources* (with $p_{00} \neq p_{10}$).

Throughout this section, $\Sigma = \{0, 1\}$ denotes the binary alphabet and $P = (p_{ij})_{i,j \in \Sigma}$ is a fixed transition matrix that satisfies

$$p_{ij} > 0 \text{ for all } i, j \in \Sigma.$$

Let $Y_n^\mu(\ell)$ be the number of *Bucket Operations* performed by *Radix Select* when searching for the ℓ-th smallest element in a list of n independent and identically distributed strings generated by a *Markov Source* with initial distribution $\mu = \mu_0 \delta_0 + \mu_1 \delta_1$ and transition matrix P.

The recursive behavior of the algorithm leads to a distributional recursion for $(Y_n^\mu(\ell))_{\ell \in \{1, \ldots, n\}}$ that is similar to the recursion on *Radix Sort* given in (1.5) on page 6:

Note that the algorithm performs n *Bucket Operations* in the first split of the list and recursively continues searching in either the left or the right sublist depending on whether ℓ is at most the

5.2. THE QUANTILE MODEL

size of the left sublist or not. Let K_n^μ denote the size of the left sublist. Also note that if $\ell > K_n^\mu$, the algorithm searches the right sublist for the element with rank $\ell - K_n^\mu$. Conditioned on K_n^μ, both sublists are independent and correspond to lists generated by *Markov Sources* with initial distribution $p_{00}\delta_0 + p_{01}\delta_1$ and $p_{10}\delta_0 + p_{11}\delta_1$ respectively (cf. the analysis of *Radix Sort* in section 1.3).

This leads to the following distributional recursion on $Y_n^\mu := (Y_n^\mu(\ell))_{\ell \in \{1,\ldots,n\}}$ for any $n \geq 2$:

$$Y_n^\mu \stackrel{d}{=} \left(\mathbb{1}_{\{\ell \leq K_n^\mu\}} Y_{K_n^\mu}^0(\ell) + \mathbb{1}_{\{\ell > K_n^\mu\}} Y_{n-K_n^\mu}^1(\ell - K_n^\mu) + n \right)_{\ell \in \{1,\ldots,n\}} \quad (5.2)$$

with $(Y_n^0)_{n \geq 0}$, $(Y_n^1)_{n \geq 0}$ and K_n^μ independent, $Y_n^i \stackrel{d}{=} (Y_n^{p_{i0}\delta_0 + p_{i1}\delta_1}(\ell))_{\ell \in \{1,\ldots,n\}}$ for both $i \in \Sigma$ and $\mathcal{L}(K_n^\mu) = B(n, \mu_0)$.

In particular, Y_n^0 and Y_n^1 satisfy for $n \geq 2$

$$\begin{aligned} Y_n^0 &\stackrel{d}{=} \left(\mathbb{1}_{\{\ell \leq I_n^0\}} Y_{I_n^0}^0(\ell) + \mathbb{1}_{\{\ell > I_n^0\}} Y_{n-I_n^0}^1(\ell - I_n^0) + n \right)_{\ell \in \{1,\ldots,n\}} \\ Y_n^1 &\stackrel{d}{=} \left(\mathbb{1}_{\{\ell \leq I_n^1\}} Y_{I_n^1}^0(\ell) + \mathbb{1}_{\{\ell > I_n^1\}} Y_{n-I_n^1}^1(\ell - I_n^1) + n \right)_{\ell \in \{1,\ldots,n\}} \end{aligned} \quad (5.3)$$

with $(Y_n^0)_{n \geq 0}$, $(Y_n^1)_{n \geq 0}$ and (I_n^0, I_n^1) independent and $\mathcal{L}(I_n^i) = B(n, p_{i0})$ for $i \in \Sigma$.

5.2.1 Worst Case Behavior

The analysis of the *Quantile Model* requires some knowledge about the worst case behavior of *Radix Select*. The only result needed for the analysis in section 5.2.2 is a linear upper bound on the average worst case behavior in the *Markov Source Model*. Nevertheless, a simple proof relying on the *Contraction Method* yields a law of large numbers that also holds for all moments.

To this end, let $M_0^0 = M_0^1 := 0$ and

$$M_n^i := \max_{1 \leq \ell \leq n} Y_n^i(\ell), \quad i \in \Sigma, n \in \mathbb{N},$$

denote the maximal number of *Bucket Operations* performed by *Radix Select* under a *Markov Source* with initial distribution $p_{i0}\delta_0 + p_{i1}\delta_1$.

The system (5.3) implies a similar system of distributional recursions for M_n^0 and M_n^1, $n \geq 2$:

$$\begin{aligned} M_n^0 &\stackrel{d}{=} M_{I_n^0}^0 \vee M_{n-I_n^0}^1 + n, \\ M_n^1 &\stackrel{d}{=} M_{I_n^1}^0 \vee M_{n-I_n^1}^1 + n, \end{aligned} \quad (5.4)$$

with $(M_k^0)_{k \geq 0}$, $(M_k^1)_{k \geq 0}$ and (I_n^0, I_n^1) independent and $\mathcal{L}(I_n^i) = B(n, p_{i0})$ for $i \in \Sigma$. Here, $x \vee y := \max\{x, y\}$ denotes the maximum of $x, y \in \mathbb{R}$.

Consider the rescaling

$$V_n^i := \begin{cases} 0, & \text{if } n = 0, \\ \frac{M_n^i}{n}, & \text{if } n \geq 1, \end{cases}$$

and note that (5.4) implies

$$V_n^0 \stackrel{d}{=} \left(\frac{I_n^0}{n} V_{I_n^0}^0\right) \vee \left(\frac{n-I_n^0}{n} V_{n-I_n^0}^1\right) + 1,$$
$$V_n^1 \stackrel{d}{=} \left(\frac{I_n^1}{n} V_{I_n^1}^0\right) \vee \left(\frac{n-I_n^1}{n} V_{n-I_n^1}^1\right) + 1,$$
(5.5)

with $(V_k^0)_{k\geq 0}$, $(V_k^1)_{k\geq 0}$ and (I_n^0, I_n^1) independent and $\mathcal{L}(I_n^i) = B(n, p_{i0})$ for $i \in \Sigma$.
The law of large numbers suggests that limits V^0 and V^1 should satisfy

$$V^0 \stackrel{d}{=} (p_{00} V^0) \vee (p_{01} V^1) + 1,$$
$$V^1 \stackrel{d}{=} (p_{10} V^0) \vee (p_{11} V^1) + 1,$$
(5.6)

with V^0, V^1 independent. In fact, there is a deterministic solution to (5.6): Note that $(V^0, V^1) = (a, b)$ solves (5.6) for $a, b \in \mathbb{R}$ if and only if (a, b) is a fixed point to the map

$$T_2 : \mathbb{R}^2 \to \mathbb{R}^2, \quad (x, y) \mapsto \bigl((p_{00} x) \vee (p_{01} y) + 1\,,\, (p_{10} x) \vee (p_{11} y) + 1\bigr). \tag{5.7}$$

Thus, the existence of a (deterministic) solution may be deduced from an analysis of T_2. More precisely, it is sufficient to show that T_2 is a contraction with respect to a complete metric on $\mathbb{R}^2$. This holds for the metric induced by the maximum norm $\|\cdot\|_\infty$ and is deduced from the next lemma.

Lemma 5.2.1. *For all $p \geq 1$ and $a, b, c, d \in \mathbb{R}$,*

$$|a \vee b - c \vee d|^p \leq |a - c|^p + |b - d|^p.$$

Here, $x \vee y := \max\{x, y\}$ denotes the maximum of $x, y \in \mathbb{R}$.

Proof. By symmetry, one may assume without loss of generality that $a \geq b$.
If $c \geq d$, the assertion holds trivially. Otherwise, it remains to show

$$|a - d|^p \leq |a - c|^p + |b - d|^p, \quad a \geq b, \ d > c.$$

This upper bound may be seen by the case analysis

$$|a - d|^p = \begin{cases} (a-d)^p \leq (a-c)^p = |a-c|^p, & \text{if } a \geq d, \\ (d-a)^p \leq (d-b)^p = |b-d|^p, & \text{if } a < d \end{cases}$$

which implies the assertion. □

Corollary 5.2.2. *Let $P = (p_{ij})_{i,j \in \Sigma}$ be a transition matrix that satisfies*

$$\max\{p_{i,j} : i, j \in \Sigma\} < 1.$$

Then, T_2 is a contraction on $(\mathbb{R}^2, \|\cdot\|_\infty)$. In particular, T_2 has a unique fixed point.

Proof. Lemma 5.2.1 implies for any $a, b, c, d \in \mathbb{R}$,

$$\|T_2(a,b) - T_2(c,d)\|_\infty^2 = |(p_{00}a) \vee (p_{01}b) - (p_{00}c) \vee (p_{01}d)|^2 \vee |(p_{10}a) \vee (p_{11}b) - (p_{10}c) \vee (p_{11}d)|^2$$
$$\leq \bigl((p_{00}^2 + p_{01}^2) \vee (p_{10}^2 + p_{11}^2)\bigr) \|(a-c, b-d)\|_\infty^2.$$

5.2. THE QUANTILE MODEL

Hence, T_2 is a contraction since $\max\{p_{i,j} : i, j \in \Sigma\} < 1$ and $p_{00} + p_{01} = 1 = p_{10} + p_{11}$.

Moreover, T_2 has a unique fixed point by Banach's fixed point theorem and the completeness of $(\mathbb{R}^2, \|\cdot\|_\infty)$. □

An application of the *Contraction Method* leads to the following result for the worst case behavior:

Theorem 5.2.3. *Let M_n^i, $i \in \Sigma$, denote the maximal number of Bucket operations performed by Radix Select when searching for any rank among n independent strings generated by a Markov Source with initial distribution $p_{i0}\delta_0 + p_{i1}\delta_1$ and transition matrix $P = (p_{kl})_{k,l \in \Sigma}$. Assume that P satisfies*

$$\max\{p_{i,j} : i, j \in \Sigma\} < 1.$$

Then, as $n \to \infty$,

$$\frac{M_n^i}{n} \xrightarrow{d} \mathfrak{m}_i, \quad i \in \Sigma,$$

where $(\mathfrak{m}_0, \mathfrak{m}_1) \in \mathbb{R}^2$ denotes the unique fixed point of the map T_2 given in (5.7).

Moreover, for any $p > 0$,

$$\lim_{n \to \infty} \frac{1}{n^p} \mathbb{E}[(M_n^i)^p] = \mathfrak{m}_i^p.$$

The required linear upper bound on the average worst case is an immediate consequence of the convergence given in theorem 5.2.3:

Lemma 5.2.4. *Consider a Markov Source that satisfies the assumptions in theorem 5.2.3. Then, there exists a constant $C > 0$ such that for all $n \in \mathbb{N}$ and both $i \in \Sigma$*

$$\mathbb{E}\left[\sup_{\ell \in \{1,\ldots,n\}} Y_n^i(\ell)\right] \leq Cn.$$

Proof. Theorem 5.2.3 implies that

$$\left(\frac{1}{n}\mathbb{E}\left[\sup_{\ell \in \{1,\ldots,n\}} Y_n^i(\ell)\right]\right)_{n \geq 1}$$

is a convergent sequence for both $i \in \Sigma$. In particular, both sequences are bounded. □

Proof of theorem 5.2.3. Recall that the rescaled random variables

$$V_n^i := \begin{cases} 0, & \text{if } n = 0, \\ \frac{M_n^i}{n}, & \text{if } n \geq 1, \end{cases}$$

satisfy the system (5.5) which is

$$V_n^0 \stackrel{d}{=} \left(\frac{I_n^0}{n} V_{I_n^0}^0\right) \vee \left(\frac{n - I_n^0}{n} V_{n - I_n^0}^1\right) + 1,$$

$$V_n^1 \stackrel{d}{=} \left(\frac{I_n^1}{n} V_{I_n^1}^0\right) \vee \left(\frac{n - I_n^1}{n} V_{n - I_n^1}^1\right) + 1,$$

with $(V_k^0)_{k\geq 0}$, $(V_k^1)_{k\geq 0}$ and (I_n^0, I_n^1) independent and $\mathcal{L}(I_n^i) = B(n, p_{i0})$ for $i \in \Sigma$.
Also recall that $(\mathfrak{m}_0, \mathfrak{m}_1) \in \mathbb{R}^2$ is the unique solution to the system

$$\begin{aligned}\mathfrak{m}_0 &= (p_{00}\mathfrak{m}_0) \vee (p_{01}\mathfrak{m}_1) + 1, \\ \mathfrak{m}_1 &= (p_{10}\mathfrak{m}_0) \vee (p_{11}\mathfrak{m}_1) + 1.\end{aligned} \tag{5.8}$$

Moreover, note that it is sufficient to show for all $p > 1$, as $n \to \infty$,

$$\ell_p(V_n^i, \mathfrak{m}_i) \longrightarrow 0, \qquad i \in \Sigma, \tag{5.9}$$

where ℓ_p denotes the Wasserstein metric given in (5.1). This is due to the fact that convergence in ℓ_p implies weak convergence and convergence of the p-th moment by lemma 5.1.4 (also note that convergence in ℓ_p implies convergence in ℓ_q for $q < p$).

To this end, consider the accompanying sequences

$$\begin{aligned}Q_n^0 &:= \left(\frac{I_n^0}{n}\mathfrak{m}_0\right) \vee \left(\frac{n - I_n^0}{n}\mathfrak{m}_1\right) + 1, \\ Q_n^1 &:= \left(\frac{I_n^1}{n}\mathfrak{m}_0\right) \vee \left(\frac{n - I_n^1}{n}\mathfrak{m}_1\right) + 1.\end{aligned}$$

It is sufficient to show

(i) $\ell_p(V_n^i, Q_n^i) \longrightarrow 0$ for both $i \in \Sigma$ as $n \to \infty$,

(ii) $\ell_p(Q_n^i, \mathfrak{m}_i) \longrightarrow 0$ for both $i \in \Sigma$ as $n \to \infty$,

since (i) and (ii) combined with the triangle inequality imply (5.9).

The convergence in (ii) is a simple consequence of the strong law of large numbers and lemma 5.2.1: Equation (5.8) and the definition of Q_n^i yield

$$\begin{aligned}(\ell_p(Q_n^i, \mathfrak{m}_i))^p &= \left\|\left(\frac{I_n^i}{n}\mathfrak{m}_0\right) \vee \left(\frac{n - I_n^i}{n}\mathfrak{m}_1\right) - (p_{i0}\mathfrak{m}_0) \vee (p_{i1}\mathfrak{m}_1)\right\|_p^p \\ &\leq \mathbb{E}\left[\left|\frac{I_n^i}{n} - p_{i0}\right|^p\right]|\mathfrak{m}_0|^p + \mathbb{E}\left[\left|\frac{n - I_n^i}{n} - p_{i1}\right|^p\right]|\mathfrak{m}_1|^p.\end{aligned}$$

Hence, (ii) follows from the strong law of large numbers and the dominated convergence theorem.

In order to obtain the convergence in (i), consider the coupling of V_n^i and Q_n^i when taking the same I_n^i in the right hand side of (5.5) and in the definition of Q_n^i:

$$\begin{aligned}(\ell_p(V_n^i, Q_n^i))^p &\leq \mathbb{E}\left[\left|\left(\frac{I_n^i}{n}V_{I_n^i}^0\right) \vee \left(\frac{n - I_n^i}{n}V_{n-I_n^i}^1\right) - \left(\frac{I_n^i}{n}\mathfrak{m}_0\right) \vee \left(\frac{n - I_n^i}{n}\mathfrak{m}_1\right)\right|^p\right] \\ &\leq \mathbb{E}\left[\left(\frac{I_n^i}{n}\right)^p\left|V_{I_n^i}^0 - \mathfrak{m}_0\right|^p + \left(\frac{n - I_n^i}{n}\right)^p\left|V_{n-I_n^i}^1 - \mathfrak{m}_1\right|^p\right]\end{aligned}$$

where the second bound holds by lemma 5.2.1. Since $\mathfrak{m}_0$ and $\mathfrak{m}_1$ are deterministic, note that

$$d_i(n) := \ell_p(V_n^i, \mathfrak{m}_i) = \|V_n^i - \mathfrak{m}_i\|_p, \quad i \in \Sigma, \, n \in \mathbb{N}.$$

5.2. THE QUANTILE MODEL

The previous upper bound on $\ell_p(V_n^i, Q_n^i)$, the convergence in (ii) and the triangle inequality yield for both $i \in \Sigma$

$$d_i(n) \leq \ell_p(V_n^i, Q_n^i) + o(1)$$

$$\leq \mathbb{E}\left[\left(\frac{I_n^i}{n}\right)^p \left|V_{I_n^i}^0 - \mathfrak{m}_0\right|^p + \left(\frac{n - I_n^i}{n}\right)^p \left|V_{n-I_n^i}^1 - \mathfrak{m}_1\right|^p\right]^{1/p} + o(1).$$

Let $d(n) := d_0(n) \vee d_1(n)$. Then,

$$\mathbb{E}\left[\left|V_{I_n^i}^0 - \mathfrak{m}_0\right|^p \mid I_n^i\right] = (d_0(I_n^i))^p \leq (d(I_n^i))^p, \quad i \in \Sigma,$$

$$\mathbb{E}\left[\left|V_{n-I_n^i}^1 - \mathfrak{m}_1\right|^p \mid I_n^i\right] = (d_1(n - I_n^i))^p \leq (d(n - I_n^i))^p, \quad i \in \Sigma,$$

and therefore, for both $i \in \Sigma$,

$$d_i(n) \leq \mathbb{E}\left[\left(\frac{I_n^i}{n}\right)^p (d(I_n^i))^p + \left(\frac{n - I_n^i}{n}\right)^p (d(n - I_n^i))^p\right]^{1/p} + o(1). \tag{5.10}$$

It only remains to deduce $d(n) \to 0$ from (5.10) which is done in two steps:

(a) The sequence $(d(n))_{n \geq 0}$ is bounded,

(b) $d(n) \to 0$ as $n \to \infty$.

For the first step, note that (5.10) implies

$$d_i(n) \leq \mathbb{E}\left[\left(\frac{I_n^i}{n}\right)^p + \left(\frac{n - I_n^i}{n}\right)^p\right]^{1/p} \sup_{0 \leq k \leq n-1} d(k) + (p_{i0}^n + p_{i1}^n) d(n) + o(1).$$

Maximizing over $i \in \Sigma$ yields for $d(n)$, $n \geq 2$,

$$d(n) \leq (1 - (p_{00}^n + p_{01}^n) \vee (p_{10}^n + p_{11}^n))^{-1} \max_{i \in \Sigma} \mathbb{E}\left[\left(\frac{I_n^i}{n}\right)^p + \left(\frac{n - I_n^i}{n}\right)^p\right]^{1/p} \sup_{0 \leq k \leq n-1} d(k) + o(1)$$

where, as $n \to \infty$,

$$(1 - (p_{00}^n + p_{01}^n) \vee (p_{10}^n + p_{11}^n))^{-1} \max_{i \in \Sigma} \mathbb{E}\left[\left(\frac{I_n^i}{n}\right)^p + \left(\frac{n - I_n^i}{n}\right)^p\right]^{1/p} \longrightarrow \max_{i \in \Sigma}(p_{i0}^p + p_{i1}^p)^{1/p} < 1.$$

Thus, an induction on n gives an upper bound for $(d(n))_{n \geq 0}$.

The upper bound implies that both

$$\eta := \sup_{n \in \mathbb{N}_0} d(n), \qquad \lambda := \limsup_{n \to \infty} d(n)$$

are finite. For part (b) let $\varepsilon > 0$ be an arbitrary small constant and $n_0 = n_0(\varepsilon) \in \mathbb{N}$ be chosen in such a way that

$$d(n) \leq \lambda + \varepsilon \text{ for all } n \geq n_0.$$

Then, since $\mathbb{P}(I_n^i \notin [n_0, n - n_0]) \to 0$ and $(d(n))_{n \geq 0}$ is bounded, (5.10) implies

$$d_i(n) \leq \mathbb{E}\left[\left(\frac{I_n^i}{n}\right)^p + \left(\frac{n - I_n^i}{n}\right)^p\right]^{1/p} (\lambda + \varepsilon) + o(1), \quad n > 2n_0, \, i \in \Sigma.$$

Maximizing over $i \in \Sigma$ and letting $n \to \infty$ yield

$$\lambda \leq \max_{i \in \Sigma}(p_{i0}^p + p_{i1}^p)^{1/p}(\lambda + \varepsilon)$$

which implies $\lambda = 0$ because $\varepsilon > 0$ may be chosen arbitrarily small and

$$\max_{i \in \Sigma}(p_{i0}^p + p_{i1}^p)^{1/p} < 1$$

for all $p > 1$ and all transition matrices with $p_{ij} \in (0,1)$. Thus,

$$\limsup_{n \to \infty} \ell_p(V_n^i, \mathfrak{m}_i) = 0$$

and the assertion follows. □

Remark 5.2.5. *The next section provides an asymptotic result on the average number of Bucket Operations depending on the rank $\ell = \lfloor tn \rfloor + 1$, $t \in [0,1]$. If $Y_n^i(\ell)$ denotes the number of Bucket Operations when selecting rank ℓ among n strings generated by a Markov Source with initial distribution $p_{i0}\delta_0 + p_{i1}\delta_1$, it is shown that, as $n \to \infty$,*

$$\frac{\mathbb{E}[Y_n^i(\lfloor tn \rfloor + 1)]}{n} \longrightarrow m_i(t), \quad t \in [0,1], \ i \in \Sigma,$$

where $m_i : [0,1] \to \mathbb{R}_0^+$, $i \in \Sigma$, are functions that satisfy for all $t \in [0,1] \setminus \{p_{i0}\}$,

$$m_i(t) = \mathbb{1}_{[0,p_{i0})}(t)p_{i0}m_0\left(\frac{t}{p_{i0}}\right) + \mathbb{1}_{(p_{i0},1]}(t)p_{i1}m_1\left(\frac{t - p_{i0}}{p_{i1}}\right) + 1.$$

Moreover $m_i(p_{i0}) = (\lim_{t \uparrow p_{i0}} m_i(t) + \lim_{t \downarrow p_{i0}} m_i(t))/2$, and thus, $\mathfrak{s}_i := \sup_{t \in [0,1]} m_i(t)$ satisfy

$$\mathfrak{s}_i = \sup_{t \in [0,1] \setminus \{p_{i0}\}} m_i(t) = (p_{i0}\mathfrak{s}_0) \vee (p_{i1}\mathfrak{s}_1) + 1, \quad i \in \Sigma.$$

Therefore, $(\mathfrak{s}_0, \mathfrak{s}_1)$ is the unique fixed point of the map T_2 given in (5.7) and the constants of theorem 5.2.3 are given by

$$\mathfrak{m}_i = \sup_{t \in [0,1]} m_i(t), \quad i \in \Sigma.$$

5.2.2 Selection of Quantiles in the Markov Source Model

This section provides a discussion on the expected number of *Bucket Operations* performed by *Radix Select* when searching for an element of rank $\lfloor tn \rfloor + 1$, $t \in [0,1)$, among n independent and identically distributed strings (on $\Sigma = \{0,1\}$) generated by a *Markov Source* with some initial distribution $\mu = \mu_0\delta_0 + \mu_1\delta_1$ and transition matrix $P = (p_{ij})_{i,j \in \Sigma}$.

Starting with the special cases $\mu = p_{i0}\delta_0 + p_{i1}\delta_1$, $i \in \Sigma$, the first order asymptotic of the expectation is derived. A similar asymptotic expansion also holds for arbitrary μ which is connected to the special cases via (5.2).

Recall that $Y_n^i(\ell)$ denotes the number of *Bucket Operations* performed by *Radix Select* when searching for an element with rank ℓ among n independent and identically distributed strings generated by a *Markov Source* with initial distribution $p_{i0}\delta_0 + p_{i1}\delta_1$.

The simplest case in the analysis of $Y_n^i(\ell)$ is the case $\ell \in \{1, n\}$:

5.2. THE QUANTILE MODEL

Lemma 5.2.6. *The expected number of Bucket Operations performed by Radix Sort satisfies, as* $n \to \infty$,

$$\mathbb{E}[Y_n^0(1)] = \frac{n}{p_{01}} + O(1), \qquad \mathbb{E}[Y_n^0(n)] = \left(\frac{p_{01}}{p_{10}} + 1\right)n + O(1),$$

$$\mathbb{E}[Y_n^1(1)] = \left(\frac{p_{10}}{p_{01}} + 1\right)n + O(1), \qquad \mathbb{E}[Y_n^1(n)] = \frac{n}{p_{10}} + O(1).$$

Proof. First note that the distributional equations (5.3) imply for any $n \geq 2$

$$Y_n^0(1) \stackrel{d}{=} \mathbb{1}_{\{I_n^0 > 0\}} Y_{I_n^0}^0(1) + \mathbb{1}_{\{I_n^0 = 0\}} Y_{n-I_n^0}^1(1) + n,$$

$$Y_n^1(1) \stackrel{d}{=} \mathbb{1}_{\{I_n^1 > 0\}} Y_{I_n^1}^0(1) + \mathbb{1}_{\{I_n^1 = 0\}} Y_{n-I_n^1}^1(1) + n.$$

Let $\nu_n^i(1) := \mathbb{E}[Y_n^i(1)]$ for $i \in \Sigma$. Then, conditioning on I_n^0 yields:

$$\nu_n^0(1) = \mathbb{E}[\nu_{I_n^0}^0(1)] + \mathbb{E}\left[\mathbb{1}_{\{I_n^0 = 0\}}(\nu_{n-I_n^0}^1(1) - \nu_{I_n^0}^0(1))\right] + n.$$

Note that lemma 5.2.4 implies

$$\left|\mathbb{E}\left[\mathbb{1}_{\{I_n^0 = 0\}}\left(\nu_{n-I_n^0}^1(1) - \nu_{I_n^0}^0(1)\right)\right]\right| \leq \mathbb{E}\left[\mathbb{1}_{\{I_n^0 = 0\}}\left(C(n - I_n^0) + CI_n^0\right)\right] = Cn\mathbb{P}\left(I_n^0 = 0\right)$$

and therefore

$$\nu_n^0(1) = \mathbb{E}[\nu_{I_n^0}^0(1)] + n + \varepsilon_0(n)$$

with $|\varepsilon_0(n)| \leq Cnp_{00}^n$. Now let $a_n := \nu_n^0(1) - \frac{n}{p_{01}}$. Then, a_n satisfies

$$a_n = \mathbb{E}[a_{I_n^0}] + n + \varepsilon_0(n) - \frac{n}{p_{01}} + \frac{np_{00}}{p_{01}} = \mathbb{E}[a_{I_n^0}] + \varepsilon_0(n)$$

which by lemma 3.1.6 implies $a_n = O(1)$ and therefore,

$$\mathbb{E}[Y_n^0(1)] = \frac{n}{p_{01}} + a_n = \frac{n}{p_{01}} + O(1) \tag{5.11}$$

which is the assertion for $\mathbb{E}[Y_n^0(1)]$. Moreover, the equality

$$\nu_n^1(1) = \mathbb{E}[\nu_n^0(I_n^1)] + \mathbb{E}\left[\mathbb{1}_{\{I_n^1 = 0\}}\left(\nu_{n-I_n^1}^1(1) - \nu_{I_n^1}^0(1)\right)\right] + n$$

yields the assertion for $\mathbb{E}[Y_n^1(1)]$ by (5.11) and $\left|\mathbb{E}\left[\mathbb{1}_{\{I_n^1 = 0\}}\left(\nu_{n-I_n^1}^1(1) - \nu_{I_n^1}^0(1)\right)\right]\right| \leq Cnp_{10}^n$.

Finally, note that

$$Y_n^0(n) \stackrel{d}{=} \mathbb{1}_{\{I_n^0 = n\}} Y_{I_n^0}^0(n) + \mathbb{1}_{\{I_n^0 < n\}} Y_{n-I_n^0}^1(n - I_n^0) + n,$$

$$Y_n^1(n) \stackrel{d}{=} \mathbb{1}_{\{I_n^1 = n\}} Y_{I_n^1}^0(n) + \mathbb{1}_{\{I_n^1 < n\}} Y_{n-I_n^1}^1(n - I_n^1) + n$$

which, by using similar arguments (replacing the role of 0 and 1 as well as I_n^i and $n - I_n^i$), yields the assertion for $\mathbb{E}[Y_n^0(n)]$ and $\mathbb{E}[Y_n^1(n)]$. □

The analysis of $Y_n^i(\lfloor tn \rfloor + 1)$ for arbitrary $t \in (0,1)$ requires some notation. Recall that the input of the algorithm is a list $\mathcal{X}_n^i := [\Xi_1, \ldots, \Xi_n]$, $i \in \Sigma$, of n independent strings $\Xi_1, \ldots, \Xi_n$ that are generated by a *Markov Source* with initial distribution $p_{i0}\delta_0 + p_{i1}\delta_1$ and transition matrix $P = (p_{kl})_{k,l \in \Sigma}$.

Let $I_i^J(n)$ denote the number of strings in $\mathcal{X}_n^i$ that have the prefix J, i.e. for $k \geq 1$ and $J = (j_1, \ldots, j_k) \in \{0,1\}^k$ let

$$I_i^J(n) = \left|\{(\xi_j)_{j \geq 1} \in \mathcal{X}_n^i : (\xi_1, \ldots, \xi_n) = J\}\right|.$$

The relation "$<$" for vectors $I = (i_1, \ldots, i_k) \in \{0,1\}^k$ and $J = (j_1, \ldots, j_k) \in \{0,1\}^k$, $k \geq 1$ is determined by the lexicographical order which is: $I < J$ if and only if there exists an integer $l \leq k$ such that

$$(i_1, \ldots, i_{l-1}) = (j_1, \ldots, j_{l-1}) \quad \text{and} \quad i_l < j_l.$$

Within this ordering, the number of strings with a prefix less than $J \in \{0,1\}^k$ is defined as

$$A_i^J(n) = \left|\{(\xi_j)_{j \geq 1} \in \mathcal{X}_n^i : (\xi_1, \ldots, \xi_k) < J\}\right|.$$

Moreover, the number of strings with a prefix which is at most $J \in \{0,1\}^k$ is defined as

$$B_i^J(n) = \left|\{(\xi_j)_{j \geq 1} \in \mathcal{X}_n^i : (\xi_1, \ldots, \xi_k) \leq J\}\right|.$$

Note that $B_i^J(n) - A_i^J(n) = I_i^J(n)$ is the number of strings with the prefix $J \in \{0,1\}^k$ and that the independence and equality in distribution among the strings in $\mathcal{X}_n^i$ imply that $B_i^J(n)$, $A_i^J(n)$ and $I_i^J(n)$ follow binomial distributions with n trials and success probabilities $p_B^i(J)$, $p_A^i(J)$ and $p_I^i(J)$ given by

$$p_I^i(J) = p_{ij_1} \prod_{\ell=2}^{k} p_{j_{\ell-1}j_\ell}, \qquad p_B^i(J) = \sum_{\widetilde{J} \leq J} p_I^i(\widetilde{J}), \qquad p_A^i(J) = p_B^i(J) - p_I^i(J).$$

Here, the second sum is taken over all $\widetilde{J} \in \{0,1\}^k$ with $\widetilde{J} \leq J$.

Note that if the element with rank ℓ has a prefix $J = (j_1, \ldots, j_m) \in \{0,1\}^m$, the m-th recursive call of *Radix Select* causes $I_i^J(n)$ *Bucket Operations* if $I_i^J(n) \geq 2$ and the algorithm terminates otherwise. Also note that the element with rank ℓ has prefix J if and only if $A_i^J(n) < \ell \leq B_i^J(n)$.

Hence, generalizing (5.3) into distributional equations after $k \geq 1$ recursive calls leads to the system

$$Y_n^i \stackrel{d}{=} \left(\sum_{J \in \{0,1\}^k} \mathbb{1}_{\{A_i^J(n) < \ell \leq B_i^J(n)\}} Y_{I_i^J(n)}^J \left(\ell - A_i^J(n)\right) \right. \tag{5.12}$$
$$\left. + n + \sum_{l=1}^{k-1} \sum_{J \in \{0,1\}^l} \mathbb{1}_{\{A_i^J(n) < \ell \leq B_i^J(n)\} \cap \{I_i^J(n) \geq 2\}} I_i^J(n) \right)_{\ell \in \{1, \ldots, n\}}$$

where $\{(Y_n^J)_{n \geq 0} : J \in \{0,1\}^k\}$ and $(I_i^J)_{J \in \{0,1\}^l, l \in \{1, \ldots, k\}}$ are independent and $\{(Y_n^J)_{n \geq 0} : J \in \{0,1\}^k\}$ is a family of independent random variables with distributions

$$\mathcal{L}\left(Y_n^{(j_1, \ldots, j_k)}\right) = \mathcal{L}\left(Y_n^{j_k}\right), \quad n \geq 0, (j_1, \ldots, j_k) \in \{0,1\}^k.$$

5.2. THE QUANTILE MODEL

Considering the quantiles $\ell = \lfloor tn \rfloor + 1$, $t \in [0, 1)$, one obtains for any $k \geq 1$ and $J = (j_1, \ldots, j_k) \in \{0, 1\}^k$

$$\{A_i^J(n) < \lfloor tn \rfloor + 1 \leq B_i^J(n)\} = \left\{\frac{A_i^J(n)}{n} \leq t < \frac{B_i^J(n)}{n}\right\}.$$

Note that the strong law of large numbers yields for any fixed $k \geq 1$ and $J = (j_1, \ldots, j_k) \in \{0, 1\}^k$

$$\frac{A_i^J(n)}{n} \longrightarrow p_A^i(J), \quad \frac{B_i^J(n)}{n} \longrightarrow p_B^i(J), \quad \text{almost surely, as } n \to \infty,$$

which leads to the deterministic intervals

$$h_i^J := \begin{cases} [p_A^i(J), p_B^i(J)), & J \in \{0,1\}^k \setminus \{(1, \ldots, 1)\} \\ [p_A^i(J), p_B^i(J)], & J = (1, \ldots, 1). \end{cases} \tag{5.13}$$

However, if $t \in (0, 1)$ hits the boundary of one of the limit intervals, i.e. if $t = p_A^i(J_0)$ for some $J_0 \in \{0, 1\}^k$, the following two events are both very likely (in fact, the probability of both events converges to $1/2$):

$$\left\{\frac{A_i^{J_0}(n)}{n} \leq t < \frac{B_i^{J_0}(n)}{n}\right\}, \quad \left\{\frac{A_i^{J_0^-}(n)}{n} \leq t < \frac{B_i^{J_0^-}(n)}{n}\right\}$$

where J_0^- is the largest vector in $\{0, 1\}^k$ with $J_0^- < J_0$ (and therefore $A_i^{J_0}(n) = B_i^{J_0^-}(n)$).

Thus, the boundary points of the limit intervals (5.13) need a special treatment. For further references, these points are denoted by

$$\mathcal{D}_n^i := \{p_A^i(J), p_B^i(J) : J \in \{0,1\}^n\} \setminus \{0, 1\}, \quad n \geq 1, \, i \in \Sigma$$

$$\mathcal{D}_\infty^i := \bigcup_{n=1}^\infty \mathcal{D}_n^i, \quad i \in \Sigma. \tag{5.14}$$

Remark 5.2.7. *There is an easy recursive construction of the limit intervals: Initialize*

$$h_i^{(0)} = [0, p_{i0}) \quad \text{and} \quad h_i^{(1)} = [p_{i0}, 1].$$

If $h_i^{(j_1, \ldots, j_k)} = [a, b)$ *for some* $a, b \in [0, 1]$ *and* $k \geq 1$, *the interval is split into*

$$h_i^{(j_1, \ldots, j_k, 0)} = [a, a + p_{j_k 0}(b - a)), \quad h_i^{(j_1, \ldots, j_k, 1)} = [a + p_{j_k 0}(b - a), b),$$

which also holds for $h_i^{(j_1, \ldots, j_k)} = [a, 1]$ *(i.e.* $(j_1, \ldots, j_k) = (1, \ldots, 1)$*) by taking*

$$h_i^{(j_1, \ldots, j_k, 0)} = [a, a + p_{j_k 0}(1 - a)), \quad h_i^{(j_1, \ldots, j_k, 1)} = [a + p_{10}(1 - a), 1].$$

Thus, $J = (j_1, \ldots, j_k)$ encodes the position of h_i^J in this splitting procedure where $j_l = 0$ corresponds to the left interval in the l-th splitting.

Let $J_k^i(t)$ be the unique vector in $\{0,1\}^k$ such that $t \in h_i^{J_k^i(t)}$. Moreover, for $t \in \mathcal{D}_k^i$, let $J_k^i(t-)$ be the largest vector in $\{0,1\}^k$ that is smaller than $J_k^i(t)$ and therefore, $t = p_B^i(J_k^i(t-))$.

Finally, let $\lambda([a,b)) = b - a$ denote the Lebesgue measure of an interval $[a,b)$ and

$$\lambda_k^i(t+) = \lambda\left(h_i^{J_k^i(t)}\right) = p_I^i(J_k^i(t)),$$

$$\lambda_k^i(t-) = \begin{cases} \lambda\left(h_i^{J_k^i(t-)}\right) = p_I^i(J_k^i(t-)), & \text{if } t \in \mathcal{D}_k^i, \\ \lambda_k^i(t+), & \text{otherwise.} \end{cases}$$

Then, the crucial parameter in the complexity of *Radix Select* in the *Quantile Model* is

$$\lambda_k^i(t) = \frac{\lambda_k^i(t+) + \lambda_k^i(t-)}{2}, \quad t \in [0,1], i \in \Sigma \tag{5.15}$$

with the convention $\lambda_0^i(t) = 1$ for all $i \in \Sigma$ and $t \in [0,1]$. Equation (5.12) leads to the following result on the asymptotic behavior of the average complexity:

Theorem 5.2.8. *The number of Bucket Operations performed by Radix Select when searching for an element of rank $\lfloor tn \rfloor + 1$ among n independent strings generated by a Markov Source with initial distribution $p_{i0}\delta_0 + p_{i1}\delta_1$ satisfies for all $t \in [0,1]$ and $i \in \Sigma$, as $n \to \infty$,*

$$\mathbb{E}[Y_n^i(\lfloor tn \rfloor + 1)] = m_i(t)n + o(n)$$

with $Y_n^i(n+1) := Y_n^i(n)$ and functions $m_i : [0,1] \to (0,\infty)$ given by

$$m_i(t) = 1 + \sum_{\ell=1}^{\infty} \lambda_\ell^i(t).$$

Here, $\lambda_\ell^i(t)$ denotes the averaged interval length defined in (5.15).

Remark 5.2.9. *The proof of theorem 5.2.8 may be extended to obtain a law of large numbers: For any $i \in \Sigma$ and $t \in (0,1) \setminus \mathcal{D}_\infty^i$*

$$\frac{Y_n^i(\lfloor tn \rfloor + 1)}{n} \xrightarrow{L_1} m_i(t), \quad n \to \infty,$$

where $\xrightarrow{L_1}$ denotes convergence in $\|\cdot\|_1$. In particular, the convergence holds in probability.

Figure 5.2 provides plots of m_0 and m_1 for some transition matrices P. Moreover, the next lemma contains several properties of m_0 and m_1:

Lemma 5.2.10. *The functions m_0 and m_1 of theorem 5.2.8 have the following properties:*

(i) The functions m_0 and m_1 are bounded. More precisely,

$$m_i(t) \leq \frac{1}{p_\wedge}, \quad i \in \Sigma, t \in [0,1]$$

where $p_\wedge := \min\{p_{kl} \ : \ k,l \in \Sigma\}$.

(ii) For both $i \in \Sigma$, m_i is continuous in t for all $t \in [0,1] \setminus \mathcal{D}_\infty^i$. Moreover, the following limits exist for all $i \in \Sigma$ and $t \in \mathcal{D}_\infty^i$:

$$m_i(t-) := \lim_{s \uparrow t} m_i(s) = 1 + \sum_{j=1}^{k_0} \lambda_k^i(t) + \lambda_{k_0}^i(t) p_{j_{k_0}^i(t)0}\left(1 + \frac{p_{01}}{p_{10}}\right)$$

$$m_i(t+) := \lim_{s \downarrow t} m_i(s) = 1 + \sum_{j=1}^{k_0} \lambda_k^i(t) + \lambda_{k_0}^i(t) p_{j_{k_0}^i(t)1}\left(1 + \frac{p_{10}}{p_{01}}\right)$$

5.2. THE QUANTILE MODEL

Figure 5.2 Plots of m_0 (red) and m_1 (blue) for different *Markov Sources*. Note that both functions tend to the affine linear function given in lemma 5.2.10 as $p_{10} \to p_{00}$.

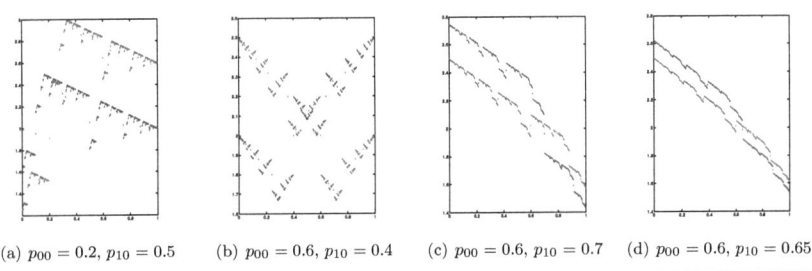

(a) $p_{00} = 0.2$, $p_{10} = 0.5$ (b) $p_{00} = 0.6$, $p_{10} = 0.4$ (c) $p_{00} = 0.6$, $p_{10} = 0.7$ (d) $p_{00} = 0.6$, $p_{10} = 0.65$

where $k_0 = k_0(i,t) \in \mathbb{N}_0$ is the smallest integer with $t \in \mathcal{D}^i_{k_0+1}$. Here, $J^i_{k_0}(t) = (j^i_1(t), \ldots j^i_{k_0}(t))$ denotes the unique vector $J \in \{0,1\}^{k_0}$ such that $t \in h^J_i$ with h^J_i given in (5.13).

In particular, if $p_{01} \neq p_{11}$ then m_i is continuous in t if and only if $t \notin \mathcal{D}^i_\infty$.

(iii) For all $i \in \Sigma$ and $t \in \mathcal{D}^i_\infty$,

$$m_i(t) = \frac{1}{2}(m_i(t-) + m_i(t+)).$$

(iv) For all $i \in \Sigma$ and $t \in [0,1] \setminus \{p_{i0}\}$,

$$m_i(t) = \mathbb{1}_{[0,p_{i0})}(t) p_{i0} m_0\left(\frac{t}{p_{i0}}\right) + \mathbb{1}_{(p_{i0},1]}(t) p_{i1} m_1\left(\frac{t - p_{i0}}{p_{i1}}\right) + 1.$$

(v) If $p_{01} = p_{11} =: p$, the function $m := m_0 = m_1$ is given by

$$m(t) = \frac{2p-1}{p(1-p)} t + \frac{1}{p}, \quad t \in [0,1].$$

The proof of lemma 5.2.10 is done at the end of the section. The result in theorem 5.2.8 also holds for arbitrary initial distributions and almost all $t \in [0,1]$:

Theorem 5.2.11. Let $Y_n^\mu(\ell)$ denote the number of Bucket Operations of Radix Select selecting a rank $1 \leq \ell \leq n$ among n independent data generated from a Markov Source with initial distribution $\mu = \mu_0 \delta_0 + \mu_1 \delta_1$ where $\mu_0 \in [0,1]$ and transition matrix $(p_{ij})_{i,j \in \{0,1\}}$ with $p_{ij} < 1$ for all $i, j = 0,1$. Moreover, let $\mathcal{D}^\mu_\infty = (\mu_0 \mathcal{D}^0_\infty) \cup (\mu_1 \mathcal{D}^1_\infty + \mu_0)$ with $\mathcal{D}^0_\infty$ and $\mathcal{D}^1_\infty$ given in (5.14). Then, the expectation of the quantiles satisfies for all $t \in [0,1] \setminus \mathcal{D}^\mu_\infty$, as $n \to \infty$,

$$\mathbb{E}[Y_n^\mu(\lfloor tn \rfloor + 1)] = m_\mu(t) n + o(n)$$

with

$$m_\mu(t) = \begin{cases} \mu_0 m_0\left(\frac{t}{\mu_0}\right) + 1, & \text{if } t < \mu_0, \\ (1-\mu_0) m_1\left(\frac{t-\mu_0}{1-\mu_0}\right) + 1, & \text{if } t > \mu_0. \end{cases}$$

The functions $m_0, m_1 : [0,1] \to (0,\infty)$ are given in theorem 5.2.8.

Proof of theorem 5.2.8. First of all, note that (5.12) implies for the rank $\ell = \lfloor tn \rfloor + 1$, $t \in [0,1)$, and for any $k \geq 2$

$$Y_n^i(\lfloor tn \rfloor + 1) \stackrel{d}{=} \sum_{J \in \{0,1\}^k} \mathbb{1}_{\left\{\frac{A_i^J(n)}{n} \leq t < \frac{B_i^J(n)}{n}\right\}} Y_{I_i^J(n)}^J \left(\lfloor tn \rfloor + 1 - A_i^J(n)\right)$$
$$+ n + \sum_{l=1}^{k-1} \sum_{J \in \{0,1\}^l} \mathbb{1}_{\left\{\frac{A_i^J(n)}{n} \leq t < \frac{B_i^J(n)}{n}\right\} \cap \{I_i^J(n) \geq 2\}} I_i^J(n) \quad (5.16)$$

with the same independence as in (5.12).

The main strategy of the proof is to choose k sufficiently large so that the first sum in (5.16) is negligible as $n \to \infty$. Moreover, the law of large numbers ensures that the remaining sum tends to $\sum_{l=1}^{k-1} \lambda_l^i(t)$.

To this end, let $\varepsilon > 0$ be an arbitrarily small constant. The proof is done by choosing k to be sufficiently large so that

(a) $S_n^{(1)} := \sum_{J \in \{0,1\}^k} \mathbb{1}_{\left\{\frac{A_i^J(n)}{n} \leq t < \frac{B_i^J(n)}{n}\right\}} Y_{I_i^J(n)}^J \left(\lfloor tn \rfloor + 1 - A_i^J(n)\right)$ satisfies

$$\frac{1}{n} \mathbb{E}[S_n^{(1)}] \leq \varepsilon \quad \text{for all sufficiently large } n.$$

(b) $S_{n,l}^{(2)} := \sum_{J \in \{0,1\}^l} \mathbb{1}_{\left\{\frac{A_i^J(n)}{n} \leq t < \frac{B_i^J(n)}{n}\right\} \cap \{I_i^J(n) \geq 2\}} I_i^J(n)$ satisfies for all $l \geq 1$

$$\left|\mathbb{E}\left[\frac{1}{n} S_{n,l}^{(2)} - \lambda_l^i(t)\right]\right| \leq \frac{\varepsilon}{k} \quad \text{for all sufficiently large } n.$$

The study of the worst case behavior in lemma 5.2.4 implies for all $J \in \{0,1\}^k$

$$\mathbb{E}\left[\mathbb{1}_{\left\{\frac{A_i^J(n)}{n} \leq t < \frac{B_i^J(n)}{n}\right\}} Y_{I_i^J(n)}^J \left(\lfloor tn \rfloor + 1 - A_i^J(n)\right)\right]$$
$$= \mathbb{E}\left[\mathbb{E}\left[\mathbb{1}_{\left\{\frac{A_i^J(n)}{n} \leq t < \frac{B_i^J(n)}{n}\right\}} Y_{I_i^J(n)}^J \left(\lfloor tn \rfloor + 1 - A_i^J(n)\right) \bigg| (I_i^{J'}(n))_{J' \in \{0,1\}^k}\right]\right]$$
$$\leq C \mathbb{E}\left[\mathbb{1}_{\left\{\frac{A_i^J(n)}{n} \leq t < \frac{B_i^J(n)}{n}\right\}} I_i^J(n)\right].$$

In particular, the expectation of the first sum in (5.16) is bounded by

$$\frac{1}{n} \mathbb{E}[S_n^{(1)}] \leq C \mathbb{E}\left[\max_{J \in \{0,1\}^k} \frac{I_i^J(n)}{n}\right].$$

The strong law of large numbers and the dominated convergence theorem yield

$$\lim_{n \to \infty} \mathbb{E}\left[\max_{J \in \{0,1\}^k} \frac{I_i^J(n)}{n}\right] = \max_{J \in \{0,1\}^k} p_I^i(J) \leq p_\vee^k.$$

5.2. THE QUANTILE MODEL

with $p_\vee := \max\{p_{lm} : l, m \in \Sigma\} < 1$. In particular, if $k \in \mathbb{N}$ is chosen in such a way that $Cp_\vee^k < \varepsilon$, an integer n_0 exists such that for all $n \geq n_0$

$$\frac{1}{n}\mathbb{E}[S_n^{(1)}] \leq \varepsilon. \tag{5.17}$$

Now let $t \in (0,1) \setminus \mathcal{D}_\infty^i$. Recall that $J_k^i(t)$ denotes the unique vector $J \in \{0,1\}^k$ so that

$$t \in [p_A^i(J), p_B^i(J)).$$

For each $\ell \in \{1, \ldots, k-1\}$, the summand in the second sum of (5.16) has the expectation

$$\frac{1}{n}\mathbb{E}[S_{n,l}^{(2)}] = p_I^i(J_l^i(t)) - \mathbb{E}\left[\mathbb{1}_{\{tn \notin [A_i^{J_l^i(t)}(n), B_i^{J_l^i(t)}(n))\} \cup \{I_i^{J_l^i(t)}(n) \leq 1\}} \frac{I_i^{J_l^i(t)}}{n}\right]$$

$$+ \frac{1}{n} \sum_{J \in \{0,1\}^l \setminus \{J_l^i(t)\}} \mathbb{E}\left[\mathbb{1}_{\{\frac{A_i^J(n)}{n} \leq t < \frac{B_i^J(n)}{n}\} \cap \{I_i^J(n) \geq 2\}} I_i^J(n)\right].$$

Since $\lambda_l^i(t) = p_I^i(J_l^i(t))$ for $t \notin \mathcal{D}_\infty^i$, one obtains the upper bound

$$\left|\frac{1}{n}\mathbb{E}[S_{n,l}^{(2)}] - \lambda_l^i(t)\right| \leq \frac{1}{n}\mathbb{P}\left(I_i^{J_l^i(t)}(n) \leq 1\right) + \mathbb{E}\left[\mathbb{1}_{\{tn \notin [A_i^{J_l^i(t)}(n), B_i^{J_l^i(t)}(n))\}} \sum_{J \in \{0,1\}^l} \frac{I_i^J(n)}{n}\right]$$

$$\leq \frac{1}{n} + \mathbb{P}\left(A_i^{J_l^i(t)}(n) > tn\right) + \mathbb{P}\left(B_i^{J_l^i(t)}(n) \leq tn\right).$$

Finally, since $t > p_A^i(J_l^i(t))$ and $t < p_B^i(J_l^i(t))$ (case $t \notin \mathcal{D}_\infty^i$), the remaining probabilities converge to zero exponentially fast (e.g. with a standard Chernoff bound given in lemma A.1.1). Thus, there exists an integer n_1 such that (b) holds for all $n \geq n_1$.

Combining equation (5.16), (a) and (b) yield for $t \in (0,1) \setminus \mathcal{D}_\infty^i$ and any $n \geq n_0 \vee n_1$

$$\left|\frac{1}{n}\mathbb{E}\left[Y_n^i(\lfloor tn \rfloor + 1)\right] - m_i(t)\right| \leq 2\varepsilon + \sum_{l=k}^{\infty} \lambda_l^i(t)$$

which, up to a suitable increase of k, is bounded by 3ε since $\lambda_l^i(t) \leq p_\vee^l$ for all $l \geq 1$. Thus, the assertion holds for all $t \in (0,1) \setminus \mathcal{D}_\infty^i$. Note that this proof may be extended to obtain

$$\lim_{n \to \infty} \mathbb{E}\left[\left|\frac{Y_n^i(\lfloor tn \rfloor + 1)}{n} - m_i(t)\right|\right] = 0, \quad t \in (0,1) \setminus \mathcal{D}_\infty^i,$$

which requires the stronger criterion

(b') $S_{n,l}^{(2)} := \sum_{J \in \{0,1\}^l} \mathbb{1}_{\{\frac{A_i^J(n)}{n} \leq t < \frac{B_i^J(n)}{n}\} \cap \{I_i^J(n) \geq 2\}} I_i^J(n)$ satisfies for all $l \geq 1$

$$\mathbb{E}\left[\left|\frac{1}{n}S_{n,l}^{(2)} - \lambda_l^i(t)\right|\right] \leq \frac{\varepsilon}{k} \quad \text{for all } n \text{ sufficiently large}.$$

However, (b') holds by similar arguments and the fact that, for any $B(n,p)$ distributed random variable $B_{n,p}$,

$$\mathbb{E}\left[\left|\frac{B_{n,p}}{n} - p\right|\right] = O\left(n^{-\frac{1}{2}}\right).$$

The assertion also holds for $t \in \{0, 1\}$ using similar arguments. However, this case is covered by the elementary proof given in lemma 5.2.6 and the fact that $J_l^i(0) = (0, \ldots, 0)$ and $J_l^i(1) = (1, \ldots, 1)$ for all $l \geq 1$.

Finally, let $t \in \mathcal{D}_\infty^i$. Let k_0 be the smallest integer such that $t \in \mathcal{D}_{k_0}^i$. Note that (a) still holds for all sufficiently large k and that (b) holds for all $l < k_0$ by applying the same proof as for $t \notin \mathcal{D}_\infty^i$. Recall that for $l \geq k_0$

$$\lambda_l^i(t) = \frac{1}{2}\lambda_l^i(t+) + \frac{1}{2}\lambda_l^i(t-) = \frac{1}{2}p_I^i(J_l^i(t)) + \frac{1}{2}p_I^i(J_l^i(t-))$$

where $J_k^i(t-)$ is the largest vector $J \in \{0,1\}^l$ with $J < J_k^i(t)$ (in lexicographical order). Thus,

$$t = p_B^i(J_k^i(t-)) = p_A^i(J_k^i(t)).$$

Once again, by a standard tail bound of the binomial distribution

$$\mathbb{P}\left(B_i^{J_k^i(t)}(n) \leq tn\right) \longrightarrow 0, \quad \mathbb{P}\left(A_i^{J_k^i(t-)}(n) \geq tn\right) \longrightarrow 0, \quad l \geq k_0,$$

which implies for $l \geq k_0$

$$\left|\frac{1}{n}\mathbb{E}[S_{n,l}^{(2)}] - \lambda_l^i(t)\right| \leq \left|\mathbb{E}\left[\mathbb{1}_{\{tn \geq A_i^{J_k^i(t)}(n)\}}\frac{I_i^{J_k^i(t)}(n)}{n}\right] - \frac{1}{2}p_A^i(J_k^i(t))\right|$$
$$+ \left|\mathbb{E}\left[\mathbb{1}_{\{tn < A_i^{J_k^i(t)}(n)\}}\frac{I_i^{J_k^i(t-)}(n)}{n}\right] - \frac{1}{2}p_B^i(J_k^i(t))\right| + o(1).$$

Hence, part (b) also holds for $l \geq k_0$ if the following holds for any $B(n, p)$ distributed random variable $B_{n,p}$ as $n \to \infty$, $p \in (0, 1)$ fixed:

$$(i) \ \mathbb{E}\left[\mathbb{1}_{\{B_{n,p} \geq np\}}\frac{B_{n,p}}{n}\right] = \frac{p}{2} + o(1), \quad (ii) \ \mathbb{E}\left[\mathbb{1}_{\{B_{n,p} < np\}}\frac{B_{n,p}}{n}\right] = \frac{p}{2} + o(1).$$

Since (ii) follows from (i) and $\mathbb{E}[B_{n,p}] = np$, it only remains to show (i) in order to finish the proof: The central limit theorem yields $\mathbb{P}(B_{n,p} \geq np) \to 1/2$ and therefore,

$$\left|\mathbb{E}\left[\mathbb{1}_{\{B_{n,p} \geq np\}}\frac{B_{n,p}}{n}\right] - \frac{p}{2}\right| = \left|\mathbb{E}\left[\mathbb{1}_{\{B_{n,p} \geq np\}}\frac{B_{n,p} - np}{n}\right]\right| + o(1) \leq \mathbb{E}\left[\left|\frac{B_{n,p} - np}{n}\right|\right] + o(1) \longrightarrow 0$$

where the convergence is justified by the law of large numbers and the dominated convergence theorem.

Thus, (a) and (b) also holds for $t \in \mathcal{D}_\infty^i$ which yields the assertion using the same arguments as for the first case. $\square$

The transfer to arbitrary initial distributions requires a more careful study of the continuity-points $t \notin \mathcal{D}_\infty^i$:

Lemma 5.2.12. *Let $(t_n)_{n \geq 0}$ be a convergent sequence in $[0, 1]$ with limit*

$$t := \lim_{n \to \infty} t_n.$$

Then, for any $i \in \Sigma$, $t \notin \mathcal{D}_\infty^i$ implies

$$\lim_{n \to \infty} \frac{1}{n}\mathbb{E}[Y_n^i(\lfloor t_n n \rfloor + 1)] = m_i(t)$$

with Y_n^i and $m_i(t)$ given in theorem 5.2.8.

5.2. THE QUANTILE MODEL

Proof. The proof is very similar to the proof of the case $t \notin \mathcal{D}_\infty^i$ in theorem 5.2.8. In fact, part (a) in the proof remains valid without any changes. For part (b) note that

$$p_A^i(J_k^i(t)) < t < p_B^i(J_k^i(t))$$

and $t_n \to t$ implies the existence of constants $\delta > 0$ and $n_0 \in \mathbb{N}$ such that

$$p_A^i(J_k^i(t)) + \delta < t_n < p_B^i(J_k^i(t)) - \delta, \quad n \geq n_0.$$

In particular, for any $l \in \{1, \ldots, k-1\}$

$$\mathbb{P}\left(t_n n \notin \left(A_i^{J_l^i(t)}(n), B_i^{J_l^i(t)}(n)\right)\right) \leq \mathbb{P}\left(t_n n \notin \left(A_k^{J_k^i(t)}(n), B_i^{J_k^i(t)}(n)\right)\right) \longrightarrow 0, \quad n \to \infty.$$

Thus, (b) remains valid for t_n and the assertion is given by the proof of theorem 5.2.8. $\square$

Proof of theorem 5.2.11. Recall that (5.2) yields for all $n \geq 2$

$$Y_n^\mu \stackrel{d}{=} \left(\mathbb{1}_{\{\ell \leq K_n^\mu\}} Y_{K_n^\mu}^0(\ell) + \mathbb{1}_{\{\ell > K_n^\mu\}} Y_{n-K_n^\mu}^1(\ell - K_n^\mu) + n\right)_{\ell \in \{1, \ldots, n\}}$$

with $(Y_n^0)_{n \geq 0}$, $(Y_n^1)_{n \geq 0}$ and K_n^μ independent and $\mathcal{L}(K_n^\mu) = B(n, \mu_0)$.
This implies for $\mu_0 = 0$ and any $t \in [0, 1]$ that

$$Y_n^\mu(t) \stackrel{d}{=} Y_n^1(t) + n$$

and the assertion is given by theorem 5.2.8 in this case. Similarly, the assertion follows for $\mu_0 = 1$ by the fact that $Y_n^\mu(t) \stackrel{d}{=} Y_n^0(t) + n$.
Now let $\mu_0 \in (0,1)$ and consider the functions $\nu_n^i : [0,1] \longrightarrow \mathbb{R}$, $n \in \mathbb{N}_0$, $i \in \Sigma$, defined as

$$\nu_n^i(t) = \frac{1}{n} \mathbb{E}[Y_n^i(\lfloor tn \rfloor \mid 1)], \quad t \subset [0,1],$$

with the convention $Y_n^i(n+1) := Y_n^i(n)$ and $\nu_0^i(t) = 0$ for all $t \in [0,1]$.
Then, (5.2) implies by conditioning on K_n^μ

$$\frac{1}{n}\mathbb{E}[Y_n^\mu(\lfloor tn \rfloor] + 1] = \mathbb{E}\left[\mathbb{1}_{\{tn < K_n^\mu\}} \frac{K_n^\mu}{n} \nu_{K_n^\mu}^0\left(\frac{tn}{K_n^\mu}\right) + \mathbb{1}_{\{tn \geq K_n^\mu\}} \frac{n - K_n^\mu}{n} \nu_{n-K_n^\mu}^1\left(\frac{tn - K_n^\mu}{n - K_n^\mu}\right)\right] + 1.$$

The strong law of large numbers and lemma 5.2.12 yield for any $t \notin \mathcal{D}_\infty^\mu$ almost surely, as $n \to \infty$,

$$\mathbb{1}_{\{tn < K_n^\mu\}} \frac{K_n^\mu}{n} \nu_{K_n^\mu}^0\left(\frac{tn}{K_n^\mu}\right) \longrightarrow \mathbb{1}_{\{t < \mu_0\}} \mu_0 m_0\left(\frac{t}{\mu_0}\right)$$

$$\mathbb{1}_{\{tn \geq K_n^\mu\}} \frac{n - K_n^\mu}{n} \nu_{n-K_n^\mu}^1\left(\frac{tn - K_n^\mu}{n - K_n^\mu}\right) \longrightarrow \mathbb{1}_{\{t \geq \mu_0\}} \mu_1 m_1\left(\frac{t - \mu_0}{\mu_1}\right).$$

Finally, the assertion follows from the dominated convergence theorem and the fact that ν_n^0 and ν_n^1 are bounded by the worst case behavior discussed in lemma 5.2.4. $\square$

We finish the section with the missing proof of lemma 5.2.10:

Proof of lemma 5.2.10. For the first part recall

$$m_i(t) = 1 + \sum_{k=1}^{\infty} \lambda_k^i(t)$$

with $\lambda_k^i(t)$ given in (5.15). Moreover, recall that

$$p_I^i(J) = p_{ij_1} \prod_{\ell=2}^{k} p_{j_{\ell-1} j_\ell} \leq p_\vee^k, \qquad J = (j_1, \ldots, j_k) \in \{0,1\}^k, \; k \geq 1,$$

with $p_\vee = \max\{p_{ij} : i, j \in \Sigma\} = 1 - p_\wedge$. This yields the upper bound

$$\lambda_k^i(t) \leq p_\vee^k, \quad k \geq 1,$$

which implies (i) by the convergence of the geometric series.

For part (ii) let $t \in [0,1] \setminus \mathcal{D}_\infty^i$ and let $\varepsilon > 0$ be an arbitrarily small constant. Then, by the proof of part (i), there exists an integer $n_0 \in \mathbb{N}$ such that for all $s \in [0,1]$

$$|m_i(s) - m_i(t)| \leq \sum_{k=1}^{n_0} |\lambda_k^i(s) - \lambda_k^i(t)| + \varepsilon. \tag{5.18}$$

Recall that $J_{n_0}^i(t) \in \{0,1\}^k$ denotes the unique vector $J \in \{0,1\}^{n_0}$ with $t \in h_i^J = [p_A^i(J), p_B^i(J))$ (or $h_i^J = [p_A^i(J), p_B^i(J)]$ if $J = (1, \ldots, 1)$) and that $t \in [0,1] \setminus \mathcal{D}_\infty^i$ implies

$$p_A^i(J_{n_0}^i(t)) < t < p_B^i(J_{n_0}^i(t)).$$

For $t \in (0,1)$ let $\delta > 0$ be chosen in such a way that

$$[t - \delta, t + \delta] \subset h_i^{J_{n_0}^i(t)}.$$

Then, for all $s \in [t - \delta, t + \delta]$ and $k \leq n_0$,

$$\lambda_k^i(s) = \lambda_k^i(t)$$

which implies $|m_i(s) - m_i(t)| \leq \varepsilon$ by (5.18). Thus, m_i is continuous in $t \in (0,1) \setminus \mathcal{D}_\infty^i$. The continuity in 0 and 1 holds by using similar arguments (pick $\delta > 0$ such that $[0, \delta] \subset h_i^{(0,\ldots,0)}$ for $t = 0$ and $[1 - \delta, 1] \subset h_i^{(1,\ldots,1)}$ for $t = 1$).

Now let $t \in \mathcal{D}_\infty^i$. Recall that k_0 is the smallest integer with $t \in \mathcal{D}_{k_0+1}^i$. In particular, t hits the boundary of the interval $h_i^{J_k^i(t)}$ if and only if $k \geq k_0 + 1$. Thus, for all s that are sufficiently close to t,

$$\lambda_k^i(s) = \lambda_k^i(t), \quad k \leq k_0.$$

On the other hand, if $k \geq k_0 + 1$,

$$t = \inf h_i^{J_k^i(t)} = \sup h_i^{J_k^i(t-)}.$$

Hence, for all $k \geq k_0 + 1$ and s sufficiently close to t,

$$s \in \begin{cases} h_i^{J_k^i(t-)}, & \text{if } s < t, \\ h_i^{J_k^i(t)}, & \text{if } s > t \end{cases}$$

5.2. THE QUANTILE MODEL

which implies with the notation in (5.15)

$$\lambda_k^i(s) = \begin{cases} \lambda_k^i(t-), & \text{if } s < t, \\ \lambda_k^i(t+), & \text{if } s > t. \end{cases}$$

Thus,

$$\lim_{s \uparrow t} m_i(s) = 1 + \sum_{j=1}^{k_0} \lambda_k^i(t) + \sum_{j=k_0+1}^{\infty} \lambda_k^i(t-),$$

$$\lim_{s \downarrow t} m_i(s) = 1 + \sum_{j=1}^{k_0} \lambda_k^i(t) + \sum_{j=k_0+1}^{\infty} \lambda_k^i(t+)$$

and therefore, (iii) holds by the definition of $\lambda_k^i(t)$. For the second part of (ii) it remains to show

$$\sum_{j=k_0+1}^{\infty} \lambda_k^i(t-) = \lambda_{k_0}^i(t) p_{j_{k_0}^i(t)0} \left(1 + \frac{p_{01}}{p_{10}}\right),$$

$$\sum_{j=k_0+1}^{\infty} \lambda_k^i(t+) = \lambda_{k_0}^i(t) p_{j_{k_0}^i(t)1} \left(1 + \frac{p_{10}}{p_{01}}\right). \tag{5.19}$$

To this end, note that, as described in the recursive representation of the intervals given in remark 5.2.7, t hits the boundary of an interval in the $k_0 + 1$-st splitting (which is the right subinterval in that splitting, causing $J_{k_0+1}^i(t) = (j_1^i(t), \ldots, j_{k_0}^i(t), 1)$ with $J_{k_0}^i(t) = (j_1^i(t), \ldots, j_{k_0}^i(t))$). After hitting the boundary of $h_i^{J_{k_0+1}^i(t)}$, t is located in the leftmost interval in every other splitting which yields

$$J_k^i(t) = (j_1^i(t), \ldots, j_{k_0}^i(t), 1, 0, \ldots, 0), \quad k > k_0 + 1.$$

Moreover, $J_k^i(t-)$ is the largest vector (in lexicographical order) which is smaller than $J_k^i(t)$. Thus,

$$J_k^i(t-) = (j_1^i(t), \ldots, j_{k_0}^i(t), 0, 1, \ldots, 1), \quad k \geq k_0 + 1$$

and (5.19) follows from the definition

$$\lambda_k^i(t+) = p_{ij_1^i(t)} \prod_{\ell=2}^{k} p_{j_{\ell-1}^i(t) j_\ell^i(t)}, \quad (j_1^i(t), \ldots, j_k^i(t)) := J_k^i(t),$$

$$\lambda_k^i(t-) = p_{ij_1^i(t-)} \prod_{\ell=2}^{k} p_{j_{\ell-1}^i(t-) j_\ell^i(t-)}, \quad (j_1^i(t-), \ldots, j_k^i(t-)) := J_k^i(t-),$$

and the limit of the geometric series. In particular, m_i is continuous in $t \in \mathcal{D}_\infty^i$ if and only if

$$\frac{p_{j_{k_0}^i(t)0}}{p_{10}} = \frac{p_{j_{k_0}^i(t)1}}{p_{01}}$$

which, for $j_{k_0}^i(t) = 0$, requires $p_{00} = p_{10}$ and, for $j_{k_0}^i(t) = 1$, requires $p_{11} = p_{01}$. Since $p_{j0} + p_{j1} = 1$ for both $j \in \Sigma$, continuity in $t \in \mathcal{D}_\infty^i$ is equivalent to $p_{00} = p_{10}$.

Part (iv) is an immediate consequence of the next property which holds for all $k \geq 1$, $i \in \Sigma$ and $t \neq p_{i0}$:

$$\lambda_k^i(t) = \begin{cases} p_{i0} \lambda_{k-1}^0 \left(\frac{t}{p_{i0}} \right), & \text{if } t < p_{i0}, \\ p_{i1} \lambda_{k-1}^1 \left(\frac{t - p_{i0}}{p_{i1}} \right), & \text{if } t > p_{i0}. \end{cases}$$

Once again, this property is best seen by considering the recursive representation of the intervals given in remark 5.2.7. If $t < p_{i0}$, t is located in the left interval of the first splitting (i.e. in the interval $[0, p_{i0})$). The relative location of t within this interval is given by t/p_{i0} and the interval is split into a left subinterval with a relative length of p_{00} (relatively to the length of the interval $[0, p_{i0})$) and a right subinterval with a relative length of p_{01}. This corresponds to the splitting of the unit interval and the case $i = 0$ and therefore,

$$\lambda_k^i(t) = p_{i0} \lambda_{k-1}^0 \left(\frac{t}{p_{i0}} \right), \quad t < p_{i0}.$$

Similar arguments also hold for the case $t > p_{i0}$.

Finally, part (v) may be deduced from (iv) and a uniqueness argument: Obviously, $m_0 = m_1$ for *Bernoulli Sources* since $p_{00} = p_{10} =: q$ and $p_{01} = p_{11} =: p$. Moreover, $m := m_0 = m_1$ satisfies by (iv)

$$m(t) = \mathbb{1}_{[0,q)}(t) q m \left(\frac{t}{q} \right) + \mathbb{1}_{[q,1]}(t) p m \left(\frac{t-q}{p} \right) + 1, \quad t \in [0,1] \setminus \{q\}. \tag{5.20}$$

It is not difficult to see that

$$f : [0,1] \to \mathbb{R}, \quad t \mapsto \frac{2p-1}{pq} t + \frac{1}{p}$$

also satisfies (5.20). Thus, for all $t \neq q$,

$$|m(t) - f(t)| \leq (q\|m - f\|_\infty) \vee (p\|m - f\|_\infty) = (p \vee q)\|m - f\|_\infty \tag{5.21}$$

where $x \vee y := \max\{x, y\}$ denotes the maximum and $\|\cdot\|_\infty$ denotes the supremum norm. Moreover, (ii) implies

$$m(q) = 1 + q \left(1 + \frac{p}{q} \right) = 2 = f(q).$$

Therefore, taking the maximum in (5.21) yields

$$\|m - f\|_\infty \leq (p \vee q)\|m - f\|_\infty$$

which implies $m = f$ for all $p \in (0,1)$. $\square$

5.2.3 A Remark on Convergence in $\mathcal{D}[0,1]$

The observations in the previous section include a law of large numbers which is

$$\frac{Y_n^i(\lfloor tn \rfloor + 1)}{n} \xrightarrow{d} m_i(t), \quad t \in [0,1] \setminus \mathcal{D}_\infty^i$$

5.2. THE QUANTILE MODEL

where $m_i : [0,1] \to \mathbb{R}$ is some bounded function. Thus, it is natural to ask if there is a proper rescaling factor $\alpha_n = o(n)$ such that the process

$$X_n^i := \left(\frac{Y_n^i(\lfloor tn \rfloor + 1) - m_i(t)n}{\alpha_n}\right)_{t \in [0,1]}, \quad n \in \mathbb{N},$$

converges in distribution in $(\mathcal{D}[0,1], d_{sk})$ to a non-degenerate limit. Here, $\mathcal{D}[0,1]$ is the space of all càdlàg functions on $[0,1]$ and d_{sk} denotes the Skorokhod distance on $\mathcal{D}[0,1]$. It is observed in [47] that this is true for *Bernoulli Sources* (i.e. the special case $p_{00} = p_{10}$) and the rescaling $\alpha_n = \sqrt{n}$. In this case, the rescaled process X_n^i converges in distribution to a centered Gaussian process.

However, it turns out that such a convergence does not hold for the case $p_{00} \neq p_{10}$. More precisely, the process

$$X_n^i := \left(X_n^i(t)\right)_{t \in [0,1]} := \left(\frac{Y_n^i(\lfloor tn \rfloor + 1) - m_i(t)n}{\sqrt{n}}\right)_{t \in [0,1]}, \quad n \in \mathbb{N}, \quad (5.22)$$

(with the convention $Y_n^i(n+1) := Y_n^i(n)$) cannot converge in distribution for $p_{00} \neq p_{10}$ due to the next theorem:

Theorem 5.2.13. *Consider a Markov Source with $p_{00} \neq p_{10}$. Then, for both $i \in \Sigma$, the family $\{\|X_n^i\|_\infty : n \in \mathbb{N}\}$ is not tight where X_n^i denotes the process defined in (5.22). In particular, the processes $(X_n^i)_{n \geq 0}$, $i \in \Sigma$, do not converge in distribution in $(\mathcal{D}[0,1], d_{sk})$.*

Proof. First of all, recall that Y_n^0 and Y_n^1 satisfy the system (5.3) which is

$$Y_n^0 \stackrel{d}{=} \left(\mathbb{1}_{\{\ell \leq I_n^0\}} Y_{I_n^0}^0(\ell) + \mathbb{1}_{\{\ell > I_n^0\}} Y_{n-I_n^0}^1(\ell - I_n^0) + n\right)_{\ell \in \{1,\ldots,n\}}$$

$$Y_n^1 \stackrel{d}{=} \left(\mathbb{1}_{\{\ell \leq I_n^1\}} Y_{I_n^1}^0(\ell) + \mathbb{1}_{\{\ell > I_n^1\}} Y_{n-I_n^1}^1(\ell - I_n^1) + n\right)_{\ell \in \{1,\ldots,n\}}$$

with $(Y_n^0)_{n>0}$, $(Y_n^1)_{n>0}$ and (I_n^0, I_n^1) independent and $\mathcal{L}(I_n^i) = B(n, p_{i0})$ for $i \in \Sigma$.
Now (5.3) implies similar equations for the rescaled quantities X_n^i, $i \in \Sigma$:

$$X_n^i \stackrel{d}{=} \left(\mathbb{1}_{\{t < I_n^i/n\}} \sqrt{\frac{I_n^i}{n}} X_{I_n^i}^0 \left(\frac{t}{I_n^i/n}\right) + \mathbb{1}_{\{t \geq I_n^i/n\}} \sqrt{\frac{n - I_n^i}{n}} X_{n-I_n^i}^1 \left(\frac{t - I_n^i/n}{1 - I_n^i/n}\right) + \tau_n^i(t)\right)_{t \in [0,1]}$$

with $(X_k^0)_{k \geq 0}$, $(X_k^1)_{k \geq 0}$, I_n^i independent and

$$\tau_n^i(t) = \frac{n - m_i(t)n + \mathbb{1}_{\{t < I_n^i/n\}} m_0\left(\frac{t}{I_n^i/n}\right) I_n^i + \mathbb{1}_{\{t \geq I_n^i/n\}} m_1\left(\frac{t - I_n^i/n}{1 - I_n^i/n}\right)(n - I_n^i)}{\sqrt{n}}, \quad t \in [0,1].$$

Now assume that $\{\|X_n^0\|_\infty : n \in \mathbb{N}\}$ was tight. Then, for $\varepsilon \in (0, 1/2)$ there exists a constant $K > 0$ such that $\mathbb{P}(\|X_n^0\|_\infty > K) \leq \varepsilon$ for all $n \in \mathbb{N}$. The distributional recursion for X_n^0 yields

$$\mathbb{P}(\|X_n^0\|_\infty > K)$$

$$= \mathbb{P}\left(\max_{t \in [0,1]} \left|\mathbb{1}_{\{t < I_n^0/n\}} \sqrt{\frac{I_n^0}{n}} X_{I_n^0}^0 \left(\frac{t}{I_n^0/n}\right) + \mathbb{1}_{\{t \geq I_n^0/n\}} \sqrt{\frac{n - I_n^0}{n}} X_{n-I_n^0}^1 \left(\frac{t - I_n^0/n}{1 - I_n^0/n}\right) + \tau_n^0(t)\right| > K\right)$$

$$\geq \mathbb{P}\left(\|\tau_n^0 \mathbb{1}_{[0, I_n^0/n]}\|_\infty > 2K, \|X_{I_n^0}^0\|_\infty \leq K\right)$$

where the last inequality is justified by a reduction of the maximum to the regime $t \in [0, I_n^0/n]$ and by the triangle inequality. Moreover, the tightness assumption implies

$$\mathbb{P}(\|X_n^0\|_\infty > K) \geq \mathbb{P}\left(\|\tau_n^0 \mathbf{1}_{[0, I_n^0/n)}\|_\infty > 2K, \|X_{I_n^0}^0\|_\infty \leq K\right) \geq \mathbb{P}\left(\|\tau_n^0 \mathbf{1}_{[0, I_n^0/n)}\|_\infty > 2K\right) - \varepsilon.$$

Thus, $\mathbb{P}\left(\|\tau_n^0 \mathbf{1}_{[0, I_n^0/n)}\|_\infty > 2K\right)$ has to be bounded by 2ε. However, the definition of τ_n^0 yields on $\{t < I_n^0/n\}$ for $t < p_{00}$:

$$\tau_n^0(t) = \frac{n - m_0(t)n + m_0\left(\frac{t}{I_n^0/n}\right) I_n^0}{\sqrt{n}}$$

$$= \frac{m_0\left(\frac{t}{I_n^0/n}\right) I_n^0 - p_{00} m_0\left(\frac{t}{p_{00}}\right) n}{\sqrt{n}} \quad \text{(lemma 5.2.10 (iv))}$$

$$= \frac{I_n^0 - n p_{00}}{\sqrt{n}} m_0\left(\frac{t}{I_n^0/n}\right) + \sqrt{n} p_{00}\left(m_0\left(\frac{t}{I_n^0/n}\right) - m_0\left(\frac{t}{p_{00}}\right)\right).$$

Skorokhod's representation theorem gives an embedding of $(I_n^0)_{n \geq 0}$ into a common probability space such that

$$\frac{I_n^0 - n p_{00}}{\sqrt{n}} \longrightarrow \mathcal{N}$$

where $\mathcal{N}$ follows the normal distribution $\mathcal{N}(0, p_{00}(1 - p_{00}))$. Since m_0 is bounded (lemma 5.2.10), the first summand of $\tau_n^0(t)$ is bounded. The almost sure convergence

$$m_0\left(\frac{t}{I_n^0/n}\right) = m_0\left(\frac{t}{p_{00} + (I_n^0 - n p_{00})/n}\right) \longrightarrow m_0\left(\frac{t}{p_{00}}-\right) \mathbf{1}_{\{\mathcal{N} > 0\}} + m_0\left(\frac{t}{p_{00}}+\right) \mathbf{1}_{\{\mathcal{N} < 0\}}$$

and the double jump of m_0 at $t \in \mathcal{D}_\infty^0$ (lemma 5.2.10 (ii)+(iii)) imply almost surely (within the Skorokhod representation)

$$\|\tau_n^0 \mathbf{1}_{[0, I_n^0/n)}\|_\infty \longrightarrow \infty.$$

Hence, $\mathbb{P}(\|\tau_n^0 \mathbf{1}_{[0, I_n^0/n)}\|_\infty > 2K) \to 1$ for any $K > 0$ contradicting

$$\varepsilon \geq \mathbb{P}(\|X_n^0\|_\infty > K) \geq \mathbb{P}\left(\|\tau_n^0 \mathbf{1}_{[0, I_n^0/n)}\|_\infty > 2K\right) - \varepsilon, \quad n \in \mathbb{N},$$

for $\varepsilon < 1/2$. Therefore, $(\|X_n^0\|_\infty)_{n \geq 0}$ is not tight. Similar arguments reveal that $(\|X_n^1\|_\infty)_{n \geq 0}$ is not tight either. □

Remark 5.2.14. *The proof of theorem 5.2.13 may be generalized to hold for any sequence*

$$\widetilde{X}_n^i(t) := \frac{Y_n^i(\lfloor tn \rfloor + 1) - m_i(t) n}{\alpha_n}, \quad t \in [0, 1],$$

with $\alpha_n = o(n)$.

5.3 Grand Averages

Recall that $Y_n^\mu(\ell)$ denotes the number of *Bucket Operations* performed by *Radix Select* when selecting the element of rank ℓ among n independent strings generated by a *Markov Source* with initial distribution $\mu = \mu_0 \delta_0 + \mu_1 \delta_1$ and transition matrix $P = (p_{ij})_{i,j \in \Sigma}$.

5.3. GRAND AVERAGES

The *Grand Averages Model* considers the sequence $(W_n^\mu)_{n\geq 0}$ given by $W_0^\mu := 0$ and

$$W_n^\mu = Y_n^\mu(U_n), \qquad n \geq 1, \tag{5.23}$$

with $(Y_n^\mu(\ell))_{\ell \in \{1,\dots,n\}}$ and U_n independent, $\mathcal{L}(U_n) = unif\{1,\dots,n\}$.

As in the other section, the analysis is based on a system of distributional recursions. First of all, note that the distributional recursion (5.2) implies

$$W_n^\mu \stackrel{d}{=} \mathbb{1}_{\{U_n \leq K_n^\mu\}} Y_{K_n^\mu}^0(U_n) + \mathbb{1}_{\{U_n > K_n^\mu\}} Y_{n-K_n^\mu}^1(U_n - K_n^\mu) + n$$

with Y_n^0, Y_n^1, K_n^μ and U_n independent.

This distributional equation may be simplified since for any $\ell \in \{1, \dots, n\}$

$$\mathcal{L}(U_n | U_n \leq \ell) = unif\{1, \dots, \ell\}$$

where $\mathcal{L}(U_n | U_n \leq \ell)$ denotes the conditional distribution of U_n on $\{U_n \leq \ell\}$.

Hence, conditioned on K_n^μ and $U_n \leq K_n^\mu$, the distribution of $Y_{K_n^\mu}^0(U_n)$ equals the distribution of $W_{K_n^\mu}^0$ where $(W_\ell^0)_{\ell \in \{0,\dots,n\}}$ is independent of K_n^μ and $\mathcal{L}(W_\ell^0) = \mathcal{L}(W_\ell^{p_{00}\delta_0 + p_{01}\delta_1})$.

Similarly, conditioned on K_n^μ and $U_n > K_n^\mu$, the distribution of $Y_{n-K_n^\mu}^1(U_n - K_n^\mu)$ equals the distribution of $W_{n-K_n^\mu}^1$ where $(W_\ell^1)_{\ell \geq 0}$ is a sequence that is independent of K_n^μ and has the distribution $\mathcal{L}(W_\ell^1) = \mathcal{L}(W_\ell^{p_{10}\delta_0 + p_{11}\delta_1})$.

Therefore, W_n^μ satisfies the following distributional recursion:

$$W_n^\mu \stackrel{d}{=} \mathbb{1}_{\{U_n \leq K_n^\mu\}} W_{K_n^\mu}^0 + \mathbb{1}_{\{U_n > K_n^\mu\}} W_{n-K_n^\mu}^1 + n, \quad n \geq 2, \tag{5.24}$$

with $(W_\ell^0)_{\ell \in \{0,\dots,n\}}$, $(W_\ell^1)_{\ell \in \{0,\dots,n\}}$, K_n^μ, U_n independent and $\mathcal{L}(W_\ell^i) = \mathcal{L}(W_\ell^{p_{i0}\delta_0 + p_{i1}\delta_1})$ for $i \in \Sigma$. In particular, W_n^0 and W_n^1 satisfy for $n \geq 2$

$$\begin{aligned} W_n^0 &\stackrel{d}{=} \mathbb{1}_{\{U_n \leq I_n^0\}} W_{I_n^0}^0 + \mathbb{1}_{\{U_n > I_n^0\}} W_{n-I_n^0}^1 + n, \\ W_n^1 &\stackrel{d}{=} \mathbb{1}_{\{U_n \leq I_n^1\}} W_{I_n^1}^0 + \mathbb{1}_{\{U_n > I_n^1\}} W_{n-I_n^1}^1 + n \end{aligned} \tag{5.25}$$

with $(W_\ell^0)_{\ell \in \{0,\dots,n\}}$, $(W_\ell^1)_{\ell \in \{0,\dots,n\}}$, U_n and (I_n^0, I_n^1) independent and $\mathcal{L}(I_n^i) = B(n, p_{i0})$ for both $i \in \Sigma$.

5.3.1 Transfers from the Quantile Model

The analysis of the *Quantile Model* immediately yields a weak convergence result for the *Grand Averages Model* given by the next lemma:

Lemma 5.3.1. Let $\{(Z_n(t))_{t \in [0,1]} : n \geq 1\}$ be a family of real valued random processes and $(Z(t))_{t \in [0,1]}$ be some limit process such that, as $n \to \infty$,

$$Z_n(t) \xrightarrow{\mathbb{P}} Z(t), \quad \text{for all } t \in [0,1] \setminus A$$

where $A \subset [0,1]$ is some set and $\xrightarrow{\mathbb{P}}$ denotes convergence in probability. Moreover, let U be uniformly distributed on $[0,1]$ and independent of $\{(Z(t))_{t \in [0,1]}, (Z_n(t))_{t \in [0,1]} : n \geq 1\}$.

Then, $\mathbb{P}(U \in A) = 0$ implies, as $n \to \infty$,
$$Z_n(U) \xrightarrow{\mathbb{P}} Z(U).$$

Proof. Let $\varepsilon > 0$ be an arbitrarily small constant. The independence between U and the processes implies
$$\mathbb{P}(|Z_n(U) - Z(U)| > \varepsilon) = \mathbb{P}(|Z_n(U) - Z(U)| > \varepsilon, U \notin A) = \int_{[0,1]\setminus A} \mathbb{P}(|Z_n(t) - Z(t)| > \varepsilon) dt$$

Now, $\mathbb{P}(|Z_n(t) - Z(t)| > \varepsilon)$ converges to 0 for all $t \in [0,1] \setminus A$ by assumption. Thus, by the dominated convergence theorem,
$$\int_{[0,1]\setminus A} \mathbb{P}(|Z_n(t) - Z(t)| > \varepsilon) dt \longrightarrow 0$$

which yields the assertion. □

Corollary 5.3.2. *The random variables $(W_n^\mu)_{n\geq 0}$ defined in (5.23) satisfy for $\mu = p_{i0}\delta_0 + p_{i1}\delta_1$, $i \in \Sigma$, as $n \to \infty$,*
$$\frac{W_n^\mu}{n} \xrightarrow{d} m_i(U)$$

where $m_i : [0,1] \to \mathbb{R}$ denotes the function given in theorem 5.2.8.

Proof. Recall that theorem 5.2.8 and remark 5.2.9 yield, as $n \to \infty$,
$$\frac{Y_n^i(\lfloor tn \rfloor + 1)}{n} \xrightarrow{\mathbb{P}} m_i(t), \quad \text{for all } t \in [0,1] \setminus \mathcal{D}_\infty^i.$$

Hence, lemma 5.3.1 yields the assertion since $W_n^{p_{i0}\delta_0 + p_{i1}\delta_1} \stackrel{d}{=} Y_n^i(\lfloor Un \rfloor + 1)$ for a uniformly on $[0,1]$ distributed U that is independent of $(Y_n^i)_{n\geq 1}$ and since $\mathcal{D}_\infty^i$ is countable which implies $\mathbb{P}(U \in \mathcal{D}_\infty^i) = 0$. □

Although weak convergence in the *Grand Averages Model* is covered by the transfer with lemma 5.3.1, there are several reasons to still study W_n^μ by using the *Contraction Method*:

- The upcoming analysis implies that the convergence in corollary 5.3.2 also holds for all moments.
- A characterization of the limit as a fixed point of some system of distributional equations provides an easy way to derive the moments of the limits.
- Convergence in the *Wasserstein* metric may easily be transferred to arbitrary initial distributions.

However, note that the fixed point equation may also be deduced from the recursion on m_i given in lemma 5.2.10 (iv) and that the transfer of weak convergence (/convergence in probability) towards arbitrary initial distributions may also be done directly. Finally, the convergence of the moments may be done by generalizing the result in the *Quantile Model* into
$$\frac{Y_n^i(\lfloor tn \rfloor + 1)}{n} \xrightarrow{L_p} m_i(t), \quad \text{for all } t \in [0,1] \setminus \mathcal{D}_\infty^i \text{ and } p > 0,$$

where L_p denotes convergence in $\|\cdot\|_p$. The proof of lemma 5.3.1 may be adapted to transfer L_p convergence by adding the following uniform bound for the dominated convergence theorem

$$\left\|\frac{Y_n^i(\lfloor tn\rfloor + 1)}{n} - m_i(t)\right\|_p \leq \frac{1}{n}\left\|Y_n^i(\lfloor tn\rfloor + 1)\right\|_p + |m_i(t)| \leq C, \quad t \in [0,1] \setminus \mathcal{D}_\infty^i$$

where the last inequality holds for some constant $C > 0$ by the worst case analysis in theorem 5.2.3 and a bound on $\sup |m_i(t)|$ given in lemma 5.2.10 (i).

5.3.2 Analysis with the Contraction Method

The main result in the *Grand Averages Model* for *Markov Sources* is a limit theorem for $(W_n^\mu/n)_{n\geq 0}$. The system (5.25) and the strong law of large numbers suggest that the limits Z^0, Z^1 of W_n^0/n and W_n^1/n should satisfy the limit system

$$\begin{aligned} Z^0 &\stackrel{d}{=} \mathbf{1}_{\{U \leq p_{00}\}} p_{00} Z^0 + \mathbf{1}_{\{U > p_{00}\}} p_{01} Z^1 + 1, \\ Z^1 &\stackrel{d}{=} \mathbf{1}_{\{U \leq p_{10}\}} p_{10} Z^0 + \mathbf{1}_{\{U > p_{10}\}} p_{11} Z^1 + 1, \end{aligned} \tag{5.26}$$

where Z^0, Z^1 and U are independent and U has the uniform distribution on $[0,1]$.

The upcoming convergence result holds in the *Wasserstein* metric ℓ_p for any $p \geq 1$. Recall that the *Wasserstein* distance for X and Y with $\mathbb{E}[|X|^p] < \infty$, $\mathbb{E}[|Y|^p] < \infty$ is defined as

$$\ell_p(X, Y) := \ell_p(\mathcal{L}(X), \mathcal{L}(Y)) = \inf\{\|W - Z\|_p : \mathcal{L}(W) = \mathcal{L}(X), \mathcal{L}(Z) = \mathcal{L}(Y)\}$$

where the infimum is taken over all random vectors (W, Z) on a common probability space with marginals $\mathcal{L}(W) = \mathcal{L}(X)$ and $\mathcal{L}(Z) = \mathcal{L}(Y)$. Here, $\|\cdot\|_p$ denotes the L_p norm which, for $p \geq 1$, is given by

$$\|W - Z\|_p = \mathbb{E}\left[|W - Z|^p\right]^{1/p}.$$

Theorem 5.3.3. *Let W_n^i, $i \in \Sigma$, denote the number of Bucket Operations of Radix Select searching for an element of a uniformly distributed rank among n independent strings generated by a Markov Source with initial distribution $p_{i0}\delta_0 + p_{i1}\delta_1$ and transition matrix $(p_{kl})_{k,l\in\{0,1\}}$ with $p_{kl} < 1$ for all $k, l \in \Sigma$.*

Then, the following convergence holds in the ℓ_p-metric for any $p \geq 1$, as $n \to \infty$:

$$\ell_p\left(\frac{W_n^i}{n}, Z^i\right) \longrightarrow 0, \quad i \in \Sigma$$

where the distributions of Z^0 and Z^1 are the unique integrable solutions of the system (5.26).

In particular, the expectations $\kappa_i := \mathbb{E}[Z^i]$, $i \in \Sigma$, are given by

$$\kappa_0 = \frac{1 + p_{01}^2 - p_{11}^2}{2(p_{00} + p_{11})(1 + p_{00}p_{11}) - 2(p_{00} + p_{11})^2},$$

$$\kappa_1 = \frac{1 + p_{10}^2 - p_{00}^2}{2(p_{00} + p_{11})(1 + p_{00}p_{11}) - 2(p_{00} + p_{11})^2}.$$

CHAPTER 5. THE RADIX SELECTION ALGORITHM

Remark 5.3.4. *Convergence in ℓ_p implies weak convergence as well as the convergence of the p-th moments. Hence, theorem 5.3.3 implies, as $n \to \infty$,*

$$\frac{W_n^i}{n} \xrightarrow{d} Z^i, \qquad \mathbb{E}\left[\left(\frac{W_n^i}{n}\right)^p\right] \longrightarrow \mathbb{E}[(Z^i)^p], \quad i \in \Sigma.$$

Proof of theorem 5.3.3. Consider the rescaled random variables $Z_0^i := 0$ and

$$Z_n^i := \frac{W_n^i}{n}, \qquad n \geq 1, \, i \in \Sigma.$$

The system (5.25) leads to a similar system for the rescaled random variables:

$$\begin{aligned}
Z_n^0 &\stackrel{d}{=} \mathbf{1}_{\{U_n \leq I_n^0\}} \frac{I_n^0}{n} Z_{I_n^0}^0 + \mathbf{1}_{\{U_n > I_n^0\}} \frac{n - I_n^0}{n} Z_{n-I_n^0}^1 + 1, \\
Z_n^1 &\stackrel{d}{=} \mathbf{1}_{\{U_n \leq I_n^1\}} \frac{I_n^1}{n} Z_{I_n^1}^0 + \mathbf{1}_{\{U_n > I_n^1\}} \frac{n - I_n^1}{n} Z_{n-I_n^1}^1 + 1,
\end{aligned} \qquad (5.27)$$

with $(Z_n^0)_{n \geq 0}$, $(Z_n^1)_{n \geq 0}$, (I_n^0, I_n^1) and U_n independent.

The strong law of large number yields for a proper realization of I_n^0, I_n^1 and $U_n = \lceil Un \rceil$ with $\mathcal{L}(U) = unif(0, 1]$ that almost surely, as $n \to \infty$,

$$\begin{aligned}
\left(\mathbf{1}_{\{U_n \leq I_n^0\}} \frac{I_n^0}{n}, \mathbf{1}_{\{U_n > I_n^0\}} \frac{n - I_n^0}{n}\right) &\longrightarrow \left(p_{00} \mathbf{1}_{\{U \leq p_{00}\}}, p_{01} \mathbf{1}_{\{U > p_{00}\}}\right), \\
\left(\mathbf{1}_{\{U_n \leq I_n^1\}} \frac{I_n^1}{n}, \mathbf{1}_{\{U_n > I_n^1\}} \frac{n - I_n^1}{n}\right) &\longrightarrow \left(p_{10} \mathbf{1}_{\{U \leq p_{10}\}}, p_{11} \mathbf{1}_{\{U > p_{10}\}}\right).
\end{aligned} \qquad (5.28)$$

Moreover, the convergence in (5.28) also holds in L_p for any $p \geq 1$ by the dominated convergence theorem.

In the spirit of the *Contraction Method*, the asymptotic behavior of the coefficients suggests that limits Z^0 and Z^1 of Z_n^0 and Z_n^1 should satisfy the system (5.26) which is given by

$$\begin{aligned}
Z^0 &\stackrel{d}{=} \mathbf{1}_{\{U \leq p_{00}\}} p_{00} Z^0 + \mathbf{1}_{\{U > p_{00}\}} p_{01} Z^1 + 1, \\
Z^1 &\stackrel{d}{=} \mathbf{1}_{\{U \leq p_{10}\}} p_{10} Z^0 + \mathbf{1}_{\{U > p_{10}\}} p_{11} Z^1 + 1,
\end{aligned}$$

with Z^0, Z^1 and U independent.

However, working with the system (5.26) requires an argument for the existence of a pair $(\mathcal{L}(Z^0), \mathcal{L}(Z^1))$ that solves (5.26). In this case, Banach's fixed point theorem and the completeness of $(\mathfrak{P}_p \times \mathfrak{P}_p, \ell_p^\vee)$ guarantees the existence of a solution to (5.26):

Consider the limit map

$$T : \mathfrak{P}_p \times \mathfrak{P}_p \longrightarrow \mathfrak{P}_p \times \mathfrak{P}_p,$$

$$\begin{pmatrix} \rho_1 \\ \rho_2 \end{pmatrix} \mapsto \begin{pmatrix} \mathcal{L}(\mathbf{1}_{\{U \leq p_{00}\}} p_{00} W + \mathbf{1}_{\{U > p_{00}\}} p_{01} Z + 1) \\ \mathcal{L}(\mathbf{1}_{\{U \leq p_{10}\}} p_{10} W + \mathbf{1}_{\{U > p_{10}\}} p_{11} Z + 1) \end{pmatrix}$$

with W, Z, U independent, $\mathcal{L}(W) = \rho_1$, $\mathcal{L}(Z) = \rho_2$ and $\mathcal{L}(U) = unif[0, 1]$.

The map T is a contraction with respect to the metric $\ell_p^\vee$. More precisely, let $\mu_1, \mu_2, \rho_1, \rho_2 \in \mathfrak{P}_p$ and consider $(W_1, W_2), (Y_1, Y_2), U$ independent such that (W_1, W_2) is an optimal ℓ_p-coupling of μ_1, ρ_1 and (Y_1, Y_2) is an optimal ℓ_p-coupling of μ_2, ρ_2.

5.3. GRAND AVERAGES

Then, $\mathbb{1}_{\{U \leq p_{i0}\}} p_{i0} W_1 + \mathbb{1}_{\{U > p_{i0}\}} p_{i1} Y_1 + 1$ and $\mathbb{1}_{\{U \leq p_{i0}\}} p_{i0} W_2 + \mathbb{1}_{\{U > p_{i0}\}} p_{i1} Y_2 + 1$ are realizations of the i-th component of $T(\mu_1, \mu_2)$ and $T(\rho_1, \rho_2)$ for $i \in \Sigma$. Thus, the *Wasserstein* distance is bounded by

$$\ell_p^\vee \left(T(\mu_1, \mu_2), T(\rho_1, \rho_2)\right) \leq \max \big\{ \|\mathbb{1}_{\{U \leq p_{00}\}} p_{00}(W_1 - W_2) + \mathbb{1}_{\{U > p_{00}\}} p_{01}(Y_1 - Y_2)\|_p,$$
$$\|\mathbb{1}_{\{U \leq p_{10}\}} p_{10}(W_1 - W_2) + \mathbb{1}_{\{U > p_{10}\}} p_{11}(Y_1 - Y_2)\|_p \big\}.$$

Moreover, the independence between U and (W_1, W_2, Y_1, Y_2) implies for both $i \in \Sigma$

$$\|\mathbb{1}_{\{U \leq p_{i0}\}} p_{i0}(W_1 - W_2) + \mathbb{1}_{\{U > p_{i0}\}} p_{i1}(Y_1 - Y_2)\|_p^p = p_{i0} \mathbb{E}\left[(p_{i0}|W_1 - W_2|)^p\right] + p_{i1} \mathbb{E}\left[(p_{i1}|Y_1 - Y_2|)^p\right]$$
$$= p_{i0}^{p+1} (\ell_p(\mu_1, \rho_1))^p + p_{i1}^{p+1} (\ell_p(\mu_2, \rho_2))^p$$
$$\leq \left(p_{i0}^{p+1} + p_{i1}^{p+1}\right) \left(\ell_p^\vee((\mu_1, \mu_2), (\rho_1, \rho_2))\right)^p.$$

This yields the upper bound

$$\ell_p^\vee \left(T(\mu_1, \mu_2), T(\rho_1, \rho_2)\right) \leq \bigg(\max\{\underbrace{p_{i0}^{p+1} + p_{i1}^{p+1}}_{< p_{i0} + p_{i1} = 1} : i \in \Sigma\} \bigg)^{1/p} \ell_p^\vee((\mu_1, \mu_2), (\rho_1, \rho_2)).$$

Hence, T is a contraction with respect to $\ell_p^\vee$ and has a unique fixed point by Banach's fixed point theorem and corollary 5.1.3.

Let $(\mathcal{L}(Z^0), \mathcal{L}(Z^1))$ be the unique fixed point. Then, Z^0 and Z^1 satisfy the system (5.26) of distributional equations.

Similar to the other proofs involving the *Contraction Method*, consider the accompanying sequences $(Q_n^0)_{n \geq 0}$ and $(Q_n^1)_{n \geq 0}$ defined as

$$Q_n^i = \mathbb{1}_{\{U_n \leq I_n^i\}} \frac{I_n^i}{n} Z^0 + \mathbb{1}_{\{U_n > I_n^i\}} \frac{n - I_n^i}{n} Z^1 + 1, \quad i \in \Sigma, \ n \geq 0,$$

with Z^0, Z^1, (I_n^0, I_n^1) and U_n being independent.

First of all, note that the asymptotic of the coefficients imply $\ell_p(Q_n^i, Z^i) \to 0$ for both $i \in \Sigma$: Consider the following coupling of $\mathcal{L}(Q_n^i)$, $\mathcal{L}(Z^i)$:

$$\mathbb{1}_{\{\lceil U n \rceil \leq I_n^i\}} \frac{I_n^i}{n} Z^0 + \mathbb{1}_{\{\lceil U n \rceil > I_n^i\}} \frac{n - I_n^i}{n} Z^1 + 1, \quad \mathbb{1}_{\{U \leq p_{i0}\}} p_{i0} Z^0 + \mathbb{1}_{\{U > p_{i0}\}} p_{i1} Z^1 + 1$$

with U, Z^0, Z^1, I_n^i being independent and distributed according to the definition of Q_n^i and Z^i.

Then, the *Wasserstein* distance is bounded by

$$\ell_p(Q_n^i, Z^i) \leq \left\| (\mathbb{1}_{\{\lceil U n \rceil \leq I_n^i\}} \frac{I_n^i}{n} - \mathbb{1}_{\{U \leq p_{i0}\}} p_{i0}) Z^0 + (\mathbb{1}_{\{\lceil U n \rceil > I_n^i\}} \frac{n - I_n^i}{n} - \mathbb{1}_{\{U > p_{i0}\}} p_{i1}) Z^1 \right\|_p$$
$$\leq \left\| \mathbb{1}_{\{\lceil U n \rceil \leq I_n^i\}} \frac{I_n^i}{n} - \mathbb{1}_{\{U \leq p_{i0}\}} p_{i0} \right\|_p \|Z^0\|_p + \left\| \mathbb{1}_{\{\lceil U n \rceil > I_n^i\}} \frac{n - I_n^i}{n} - \mathbb{1}_{\{U > p_{i0}\}} p_{i1} \right\|_p \|Z^1\|_p$$
$$\longrightarrow 0,$$

where the convergence holds by (5.28).

Hence, the triangle inequality yields

$$\ell_p(Z_n^i, Z) \leq \ell_p(Z_n^i, Q_n^i) + o(1), \qquad i \in \Sigma. \tag{5.29}$$

Now let $(Z_n^0)_{n\geq 0}$, $(Z_n^1)_{n\geq 0}$, Z^0, Z^1, (I_n^0, I_n^1) and U be independent and $(Z_n^i)_{n\geq 0}$, Z^i be an optimal ℓ_p coupling of $\{\mathcal{L}(Z_n^i), \mathcal{L}(Z^i) : n \geq 0\}$. Then, the definition of Q_n^i and the distributional recursion (5.27) yield for both $i \in \Sigma$

$$\ell_p(Z_n^i, Q_n^i) \leq \left\| \mathbf{1}_{\{U_n \leq I_n^i\}} \frac{I_n^i}{n}(Z_{I_n^i}^0 - Z^0) + \mathbf{1}_{\{U_n > I_n^i\}} \frac{n - I_n^i}{n}(Z_{n-I_n^i}^1 - Z^1) \right\|_p.$$

Note that either $\mathbf{1}_{\{U_n \leq I_n^i\}}$ or $\mathbf{1}_{\{U_n > I_n^i\}}$ equals zero and therefore,

$$\mathbb{E}\left[\left(\mathbf{1}_{\{U_n \leq I_n^i\}} \frac{I_n^i}{n}(Z_{I_n^i}^0 - Z^0) + \mathbf{1}_{\{U_n > I_n^i\}} \frac{n - I_n^i}{n}(Z_{n-I_n^i}^1 - Z^1)\right)^p\right]$$
$$= \mathbb{E}\left[\mathbf{1}_{\{U_n \leq I_n^i\}} \left(\frac{I_n^i}{n}\right)^p (Z_{I_n^i}^0 - Z^0)^p + \mathbf{1}_{\{U_n > I_n^i\}} \left(\frac{n - I_n^i}{n}\right)^p (Z_{n-I_n^i}^1 - Z^1)^p\right].$$

Let $\Delta_i(n) := \ell_p(Z_n^i, Z^i)$ for $n \geq 0$. Then, conditioning on I_n^i and U_n and applying (5.29) yield

$$\Delta_i(n) \leq \mathbb{E}\left[\mathbf{1}_{\{U_n \leq I_n^i\}} \left(\frac{I_n^i}{n}\right)^p (\Delta_0(I_n^i))^p + \mathbf{1}_{\{U_n > I_n^i\}} \left(\frac{n - I_n^i}{n}\right)^p (\Delta_1(n - I_n^i))^p\right]^{1/p} \tag{5.30}$$
$$+ o(1).$$

The proof is finished by a couple of standard arguments to show that (5.30) implies for the sequence $(\Delta(n))_{n\geq 1}$ given by $\Delta(n) := \max\{\Delta_0(n), \Delta_1(n)\}$ that $\Delta(n)$ converges to zero. The assertion follows from the definition of $\Delta(n)$.

The convergence is shown in two steps:

(a) The bound (5.30) implies that $(\Delta(n))_{n\geq 0}$ is a bounded sequence,

(b) The bounded sequence $(\Delta(n))_{n\geq 0}$ converges to zero.

To this end, the upper bound (5.30) is split into the regimes $\{I_n^i = 0\}$, $\{I_n^i = n\}$ and $\{1 \leq I_n^i \leq n-1\}$:

$$\Delta_i(n) \leq (p_{i0}^n + p_{i1}^n) \Delta(n) + \mathbb{E}\left[\mathbf{1}_{\{U_n \leq I_n^i\}} \left(\frac{I_n^i}{n}\right)^p + \mathbf{1}_{\{U_n > I_n^i\}} \left(\frac{n - I_n^i}{n}\right)^p\right]^{1/p} \max_{k \leq n-1} \Delta(k) + o(1).$$

Maximizing over $i \in \Sigma$ yields

$$\Delta(n) \leq (1 - 2p_\vee^n)^{-1} \alpha_n \max_{k \leq n-1} \Delta(k) + o(1)$$

with $\alpha_n = \max_{i \in \Sigma} \left\{ \mathbb{E}\left[\mathbf{1}_{\{U_n \leq I_n^i\}} \left(\frac{I_n^i}{n}\right)^p + \mathbf{1}_{\{U_n > I_n^i\}} \left(\frac{n - I_n^i}{n}\right)^p\right]^{1/p} \right\}$ and $p_\vee = \max_{l,m \in \Sigma} p_{lm}$.

Note that the convergence of the coefficients given in (5.28) implies that, as $n \to \infty$,

$$\alpha_n \longrightarrow \max_{i \in \Sigma} \left\{ \left(p_{i0}^{p+1} + p_{i1}^{p+1}\right)^{1/p} \right\} < 1.$$

5.3. GRAND AVERAGES

Moreover, the assumption $p_\vee < 1$ yields $(1 - 2p_\vee^n)^{-1} \to 1$. In particular, there exists an $\varepsilon > 0$ and $n_0 \in \mathbb{N}$ such that $(1 - 2p_\vee^n)^{-1} \alpha_n \leq 1 - \varepsilon$ for $n \geq n_0$ and therefore,

$$\Delta(n) \leq (1 - \varepsilon) \max_{k \leq n-1} \Delta(k) + o(1), \quad n \geq n_0.$$

With $M \geq \max_{k \leq n_0 - 1} \Delta(k)$ (and a suitable increase of n_0 such that the $o(1)$ term is bounded by εM) this yields

$$\sup_{n \in \mathbb{N}} \Delta(n) \leq M < \infty.$$

It remains to show the convergence stated in (b). To this end, let

$$\beta = \limsup_{n \to \infty} \Delta(n).$$

For any $\delta > 0$, there exists a constant $n_1 \in \mathbb{N}$ such that $\Delta(n) \leq \beta + \delta$ for all $n \geq n_1$. Splitting the upper bound (5.30) into the regimes $\{I_n^i < n_1\}$, $\{I_n^i > n - n_1\}$ and $\{n_1 \leq I_n^i \leq n - n_1\}$ yields for any $n \geq n_1$ and $i \in \Sigma$

$$\Delta_i(n) \leq \alpha_n(\beta + \delta) + M(\mathbb{P}(I_n^i < n_1) + \mathbb{P}(I_n^i > n - n_1)) + o(1)$$

with $\alpha_n = \max_{i \in \Sigma} \left\{ \mathbb{E}\left[\mathbb{1}_{\{U_n \leq I_n^i\}} \left(\frac{I_n^i}{n}\right)^p + \mathbb{1}_{\{U_n > I_n^i\}} \left(\frac{n - I_n^i}{n}\right)^p \right]^{1/p} \right\}$.

Note that, as $n \to \infty$, $\mathbb{P}(I_n^i < n_1) + \mathbb{P}(I_n^i > n - n_1) = o(1)$ for fixed n_1. Hence, by taking the maximum over $i \in \Sigma$,

$$\Delta(n) \leq \alpha_n(\beta + \delta) + o(1).$$

Recall that $\alpha_n \to \alpha < 1$ and therefore, for suitable $\tilde{n}_1 \geq n_1$ and all $n \geq \tilde{n}_1$,

$$\Delta(n) \leq (\alpha + \delta)(\beta + \delta) + o(1).$$

This implies for $\beta = \limsup_{n \to \infty} \Delta(n)$ that

$$\beta \leq (\alpha + \delta)(\beta + \delta).$$

Note that this holds for all $\delta > 0$ and thus, letting $\delta \downarrow 0$,

$$\beta \leq \alpha \beta$$

which implies $\beta = 0$ since $\alpha < 1$.

Hence, as $n \to \infty$,

$$\ell_p(Z_n^i, Z^i) = \Delta_i(n) \leq \Delta(n) \longrightarrow 0, \quad i \in \Sigma,$$

which is the desired result.

It remains to compute $\mathbb{E}[Z^0]$ and $\mathbb{E}[Z^1]$. One obtains by taking the expectation in (5.26) and by plugging in the independence of the random variables that $\kappa_i = \mathbb{E}[Z^i]$, $i \in \Sigma$, satisfy

$$\kappa_0 = p_{00}^2 \kappa_0 + p_{01}^2 \kappa_1 + 1,$$
$$\kappa_1 = p_{10}^2 \kappa_0 + p_{11}^2 \kappa_1 + 1.$$

This system of linear equations is easy to solve and leads to the results for κ_0 and κ_1 stated in the theorem. Note that higher moments of Z^0 and Z^1 may also be calculated through the use of similar arguments. □

Theorem 5.3.3 may be transfered to arbitrary initial distributions:

Theorem 5.3.5. *Let W_n^μ denote the number of Bucket Operations of Radix Select searching for an element of a uniformly distributed rank among n independent strings generated by a Markov Source with initial distribution $\mu = \mu_0 \delta_0 + \mu_1 \delta_1$, $\mu_0 \in [0,1]$, and transition matrix $(p_{ij})_{i,j \in \{0,1\}}$ with $p_{ij} < 1$ for all $i,j = 0,1$. Then, as $n \to \infty$,*

$$\frac{W_n^\mu}{n} \xrightarrow{d} Z,$$

where the convergence also holds for all moments. The distribution of Z is given by

$$Z \stackrel{d}{=} B_{\mu_0} \mu_0 Z^0 + (1 - B_{\mu_0})(1 - \mu_0) Z^1 + 1, \tag{5.31}$$

where B_{μ_0}, Z^0, Z^1 are independent, B_{μ_0} follows the Bernoulli distribution $B(\mu_0)$ and Z^0, Z^1 are the limits given in Theorem 5.3.3.

In particular, the expectation satisfies

$$\mathbb{E}[W_n^\mu] = (\mu_0^2 \kappa_0 + \mu_1^2 \kappa_1 + 1) n + o(n)$$

with κ_0 and κ_1 given in theorem 5.3.3.

Proof. By lemma 5.1.4 it is sufficient to show that for any $p \geq 1$, as $n \to \infty$,

$$\ell_p\left(\frac{W_n^\mu}{n}, Z\right) \longrightarrow 0 \tag{5.32}$$

where ℓ_p denotes the *Wasserstein* metric.

Note that this implies in particular, as $n \to \infty$,

$$\frac{1}{n}\mathbb{E}[W_n^\mu] \longrightarrow \mathbb{E}[Z] = \mathbb{E}[B_{\mu_0}\mu_0 Z^0 + (1-B_{\mu_0})(1-\mu_0)Z^1 + 1] = \mu_0^2 \kappa_0 + \mu_1^2 \kappa_1 + 1$$

where the second equality holds for $\kappa_i = \mathbb{E}[Z^i]$, $i \in \Sigma$, due to the independence between Z^0, Z^1 and B_{μ_0}.

Hence, it only remains to show (5.32). To this end, let $(Q_n^\mu)_{n \geq 0}$ be the accompanying sequence defined as

$$Q_n^\mu = \mathbf{1}_{\{U_n \leq K_n^\mu\}} \frac{K_n^\mu}{n} Z^0 + \mathbf{1}_{\{U_n > K_n^\mu\}} \frac{n - K_n^\mu}{n} Z^1 + 1$$

where U_n, K_n^μ, Z^0, Z^1 are independent, Z^0 and Z^1 denote the limits given in theorem 5.3.3, U_n is uniformly distributed on $\{1, \ldots, n\}$ and K_n^μ follows the binomial distribution $B(n, \mu_0)$.

The coupling $U_n = \lceil Un \rceil$ and $B_{\mu_0} = \mathbf{1}_{\{U \leq \mu_0\}}$, with U uniformly distributed on $[0,1]$ and independent of K_n^μ, Z^0, Z^1, yields

$$\ell_p(Q_n^\mu, Z) \leq \left\|\mathbf{1}_{\{\lceil Un \rceil \leq K_n^\mu\}}\frac{K_n^\mu}{n} - \mathbf{1}_{\{U \leq \mu_0\}}\mu_0\right\|_p \|Z^0\|_p + \left\|\mathbf{1}_{\{\lceil Un \rceil > K_n^\mu\}}\frac{n - K_n^\mu}{n} - \mathbf{1}_{\{U > \mu_0\}}\mu_1\right\|_p \|Z^1\|_p$$

with Z^0, Z^1, U and K_n^μ being independent and $\mathcal{L}(U) = unif(0,1]$. Therefore, the strong law of large numbers and the dominated convergence theorem imply, as $n \to \infty$,

$$\ell_p(Q_n^\mu, Z) \longrightarrow 0.$$

5.3. GRAND AVERAGES

In accordance with the triangle inequality, it only remains to show, as $n \to \infty$,

$$\ell_p\left(\frac{W_n^\mu}{n}, Q_n^\mu\right) \longrightarrow 0.$$

To this end, note that

$$\ell_p\left(\frac{W_n^\mu}{n}, Q_n^\mu\right) \leq \left\|\mathbf{1}_{\{U_n \leq K_n^\mu\}} \frac{K_n^\mu}{n}\left(Z_{K_n^\mu}^0 - Z^0\right)\right\|_p + \left\|\mathbf{1}_{\{U_n > K_n^\mu\}} \frac{n - K_n^\mu}{n}\left(Z_{n-K_n^\mu}^1 - Z^1\right)\right\|_p$$

$$\leq \left\|Z_{K_n^\mu}^0 - Z^0\right\|_p + \left\|Z_{n-K_n^\mu}^1 - Z^1\right\|_p.$$

Theorem 5.3.3 yields that the distances $\Delta_i(n) := \ell_p(Z_n^i/n, Z^i)$, $i \in \Sigma$, converge to zero and therefore, by conditioning on K_n^μ,

$$\left\|Z_{K_n^\mu}^0 - Z^0\right\|_p = \mathbb{E}\left[\left(Z_{K_n^\mu}^0 - Z^0\right)^p\right]^{1/p} = \mathbb{E}\left[(\Delta_0(K_n^\mu))^p\right]^{1/p} \longrightarrow 0$$

where the exchange of limit and expectation is justified by the dominated convergence theorem (note that $(\Delta_0(n))_{n \geq 0}$ is bounded) and $K_n^\mu \to \infty$ almost surely as $n \to \infty$.

Similar arguments reveal

$$\left\|Z_{n-K_n^\mu}^1 - Z^1\right\|_p \longrightarrow 0 \qquad (n \to \infty)$$

and the assertion follows. $\square$

5.3.3 A Remark on the Concentration for Grand Averages

Note that for the special case $p_{ij} = \mu_i = \frac{1}{2}$, $i,j \in \Sigma$, the *Markov Source Model* reduces to the *symmetric Bernoulli Source Model*. In this case it was shown in [15] that the complexity W_n in theorem 5.3.5 satisfies, as $n \to \infty$,

$$\frac{W_n - 2n}{\sqrt{2n}} \xrightarrow{d} \mathcal{N}(0,1). \tag{5.33}$$

Theorem 5.3.5 also applies to the *symmetric Bernoulli Source Model*: In this case, the system (5.26) is solved by the deterministic limit $Z^0 = Z^1 = 2$. Hence, the result of theorem 5.3.5 yields for a symmetric Bernoulli Source

$$\frac{W_n}{n} \xrightarrow{d} 2.$$

However, for $(p_{00}, p_{01}, p_{10}, p_{11}) \neq (\frac{1}{2}, \frac{1}{2}, \frac{1}{2}, \frac{1}{2})$ the system (5.26) has no deterministic solution and theorem 5.3.5 yields

$$\frac{W_n}{n} \xrightarrow{d} Z$$

for some limit Z with $\mathrm{Var}(Z) > 0$. Hence, W_n is less concentrated for any *Markov Source* other than the *symmetric Bernoulli Source*. The study of the *Quantile Model* provides an explanation to this phenomena:

Let $Y_n^\mu(\ell)$ denote the number of *Bucket Operations* performed by *Radix Select* when searching for an element of rank ℓ among n independent strings generated by a *Markov Source* with initial distribution μ and transition matrix P. Recall that theorem 5.2.11 yields for almost all $t \in [0, 1]$

$$\frac{1}{n}\mathbb{E}[Y_n^\mu(\lfloor tn \rfloor + 1)] \longrightarrow m_\mu(t)$$

with $m_\mu(t)$ given in theorem 5.2.11. Note that $m_\mu(t) = 2$ for all $t \in [0,1]$ if the source is a *symmetric Bernoulli Source* (i.e. $\mu = \frac{1}{2}\delta_0 + \frac{1}{2}\delta_1$ and $p_{ij} = \frac{1}{2}$ for all $i, j \in \Sigma$). For any other Markov Source, m_μ is not constant. Thus, choosing $t = U$ with a uniformly on $[0,1]$ distributed U leads to a linear fluctuation of $Y_n^\mu(\lfloor Un \rfloor + 1)$ which coincides with the fluctuation of W_n because $\lfloor Un \rfloor + 1$ is uniformly distributed on $\{1, \ldots, n\}$.

This observation coincides with the representation of the limits given in corollary 5.3.2: For any $i \in \Sigma$ and initial distribution $\mu = p_{i0}\delta_0 + p_{i1}\delta_1$,

$$\frac{Y_n^i(\lfloor Un \rfloor + 1)}{n} \xrightarrow{d} m_i(U)$$

with U uniformly distributed on $[0,1]$ and independent of $(Y_n^i(t))_{t \in [0,1], n \in \mathbb{N}}$. Therefore, the weak limits of W_n^0 and W_n^1 given in theorem 5.3.3 satisfy $\mathcal{L}(Z_i) = \mathcal{L}(m_i(U))$ which may also be extended to hold for any initial distribution.

Chapter 6
Conclusions

Mainly two types of algorithms on words were discussed in this thesis: The sorting algorithm *Radix Sort*, which is also connected to the path length in *Digital Trees*, and the selection algorithm *Radix Select*. A study of these algorithms requires a suitable (stochastic) model for the input. Most of the literature is focused on the analysis of the *Bernoulli Source Model*. This thesis is focused on the analysis of the *Markov Source Model*, which unlike the *Bernoulli Source Model* allows dependencies between two consecutive symbols in each string.

It turns out that most of the methods involved in the analysis of the *Bernoulli Source Model* may be generalized to the analysis of *Markov Sources*. In particular, the asymptotic results derived for *Radix Sort* on *Markov Sources* coincide with the results for the *asymmetric Bernoulli Source Model*. One of the major difficulties in the analysis of *Radix Sort* is the derivation of a proper asymptotic expansion of the mean, which is usually required up to the order of the standard deviation in order to apply the *Contraction Method*. Since such an expansion seems far out of reach with the current methods, a slightly different approach was developed within this thesis.

This approach is based on the fact that Lipschitz-continuity of the error term in the expansion of the mean is easy to verify but still a sufficiently powerful tool in the analysis of the variance and the application of the *Contraction Method*. In fact, this approach seems promising for other recursive structures which have splitter (subproblem sizes) that are more concentrated than the quantity itself (e.g. in *Radix Sort*, the subproblem size K_n^μ follow the binomial distribution and thus $\text{Var}(K_n^\mu) = O(n)$ whereas the number B_n^μ of *Bucket Operations* satisfies $\text{Var}(B_n^\mu) = \Theta(n \log n)$ for any *Markov Source* other than the *symmetric Bernoulli Source*). Moreover, Hölder-continuity with some Hölder-exponent $\beta < 1$ of error terms might also be helpful for other recursive structures X_n which have a splitter K_n with $\mathbb{E}[|K_n - \mathbb{E}[K_n]|^{2\beta}] = o(\text{Var}(X_n))$. However, Lipschitz-continuity of functions on $\mathbb{N}_0$ is easier to verify due to the fact that this only requires a bound on the increments.

Not all results known from the analysis of the *Bernoulli Source Model* also hold for *Markov Sources*: The analysis of *Radix Select* on *Markov Sources* reveals some features which do not appear in the analysis of *Bernoulli Sources*. In particular, the limiting behavior of the mean in the *Quantile Model* leads to a discontinuous function for *Markov Sources* which becomes an affine linear function when considering *Bernoulli Sources*. Moreover, there is no weak convergence of the rescaled process, which is known to converge to a Gaussian limit for *Bernoulli Sources*.

Although all results were only proven for the binary alphabet $\Sigma = \{0, 1\}$, a generalization to larger alphabets seems to be straightforward, at least for most of the proofs. However, some results rely on an explicit calculation involving the stationary distribution of the Markov chain (e.g. the transfer lemmas 3.6 and 3.1.4 for the first order asymptotic of mean, variance and the Lipschitz-continuity). Proving these results might become more challenging for larger alphabets but they should be fairly easy to achieve with analytical methods.

6.1 Open Problems

Mainly three aspects of algorithms were discussed in this thesis: mean, variance and weak limits. Other aspects, such as tail bounds on the performance of any of these algorithms, still need to be discussed.

Moreover, there are several open questions about *Radix Select*:

- Is there convergence in distribution of the marginals $Y_n^\mu(\lfloor tn \rfloor + 1) - m_\mu(t))/\sqrt{n}$, $t \in [0,1]$? Does the limit depend on whether m_μ is continuous in t or not? Is the limit Gaussian (as it is in the *Bernoulli Source Model*) for some/all t?

- The analysis of $M_n^\mu = \max_{\ell \in \{1,\ldots,n\}} Y_n^\mu(\ell)$ only considered the moments of M_n^μ since the expectation is relevant for the analysis of the quantile model. What is the asymptotic behavior of $\text{Var}(M_n^\mu)$? Is there a limit law for M_n^μ after rescaling?

- In the *Grand Averages Model*, a weak limit Z^i of $Y_n^i(U_n)/n$ was derived, where U_n is uniformly distributed on $\{1, \ldots, n\}$. This limit was characterized by a fixed point equation and by the representation $\mathcal{L}(Z^i) = \mathcal{L}(m_i(U))$, where $m_i(t) = \lim_{\to\infty} \mathbb{E}[Y_n^i(\lfloor tn \rfloor + 1)]/n$ and $\mathcal{L}(U) = unif[0, 1]$. The fixed point equation gave some information about the moments of Z^i and some properties of m_i may be transferred to Z^i. However, there might be some other interesting properties of Z^i which are not known so far: Is there a density f_i for the distribution of Z^i? Is this density f_i a smooth function?

Moreover, there are a few gaps that need to be filled in order to transfer the result on *Digital Search Trees* to the performance of the *Lempel-Ziv'78 Parsing Scheme*. This compression algorithm fragments the message into blocks which is described in detail in [69]. The distribution of the number of blocks starting with symbol zero becomes harder to handle for *Markov Sources* (obviously, this quantity follows the binomial distribution for *Bernoulli Sources*) which makes a transfer of the result on *Digital Search Trees* more complicated than for *Bernoulli Sources*.

Appendix

A.1 Tail Bounds for the Binomial Distribution

Most of the asymptotic results on mean and variance in this thesis rely on the concentration of the binomial distribution. A useful tail bound in many applications was derived by Chernoff in 1952:

Theorem A.1.1 (Chernoff, 1952). *Let X be a binomial $B(n,p)$ distributed random variable for $p \in (0,1)$ and $n \in \mathbb{N}$. Then, for all $\varepsilon > 0$*

$$\mathbb{P}(X - np \geq \varepsilon n) \leq \exp\left(-2n\varepsilon^2\right),$$
$$\mathbb{P}(X - np \leq -\varepsilon n) \leq \exp\left(-2n\varepsilon^2\right).$$

Note that this upper bound does not depend on p. This causes the upper bound to be very rough for p that are close to 0 or 1. A tail bound that takes the effect of p into account is given by Bernstein in 1924:

Theorem A.1.2 (Bernstein, 1924). *Let $X_1, \ldots, X_n$ be independent random variables that satisfy $\mathbb{E}[X_i] = 0$ and $|X_i| \leq c$ almost surely for some constant $c > 0$ and all $i \in \{1, \ldots, n\}$. Moreover, let $\sigma^2 = \frac{1}{n}\sum_{i=1}^{n} \text{Var}(X_i)$. Then, for all $\varepsilon > 0$,*

$$\mathbb{P}\left(\sum_{i=1}^{n} X_i \geq \varepsilon n\right) \leq \exp\left(-\frac{n\varepsilon^2}{2\sigma^2 + 2c\varepsilon/3}\right).$$

Bernstein's inequality implies the following tail bounds for the binomial distribution:

Corollary A.1.3. *Let X be a binomial $B(n,p)$ distributed random variable for $p \in (0,1)$ and $n \in \mathbb{N}$. Then, for all $\varepsilon > 0$*

$$\mathbb{P}(X \geq (1+\varepsilon)\mathbb{E}[X]) \leq \exp\left(-\frac{n\varepsilon^2 p}{2(1-p) + 2\varepsilon/3}\right),$$
$$\mathbb{P}(X \leq (1-\varepsilon)\mathbb{E}[X]) \leq \exp\left(-\frac{n\varepsilon^2 p}{2(1-p) + 2\varepsilon/3}\right).$$

Proof. Let $B_1, \ldots, B_n$ be independent Bernoulli $B(p)$ distributed random variables and consider the centered random variables $X_i := B_i - p$ for $i = 1, \ldots, n$. Theorem A.1.2 implies (with $c = 1$)

$$\mathbb{P}(X \geq (1+\varepsilon)\mathbb{E}[X]) = \mathbb{P}\left(\sum_{i=1}^n X_i \geq \varepsilon p n\right)$$
$$\leq \exp\left(-\frac{n\varepsilon^2 p^2}{2p(1-p) + 2\varepsilon p/3}\right)$$
$$= \exp\left(-\frac{n\varepsilon^2 p}{2(1-p) + 2\varepsilon/3}\right).$$

For the second bound note that

$$\mathbb{P}(X \leq (1-\varepsilon)\mathbb{E}[X]) = \mathbb{P}\left(n - X \geq \left(1 + \frac{\varepsilon p}{1-p}\right)\mathbb{E}[n-X]\right)$$

which yields the assertion by the first bound and the fact that $n - X$ follows the binomial distribution $B(n, 1-p)$. □

A.2 Moment Bounds

The concentration of the binomial distribution causes several moments in the calculation of mean and variance to be asymptotically negligible. For the sake of completeness, there are full proofs for all upper bound given in this section. Recall that $x \log x := 0$ for $x = 0$.

Lemma A.2.1. *Let $p \in (0,1)$ be some constant and $B_{n,p}$ be binomial $B(n,p)$ distributed for $n \in \mathbb{N}$. Then, as $n \to \infty$,*

$$\mathrm{Var}\left(B_{n,p}(\log(B_{n,p}/n) - \log p) + (n - B_{n,p})(\log(1 - B_{n,p}/n) - \log(1-p))\right) = O(\log n).$$

Proof. Consider the function $\phi_p : [0,1] \to \mathbb{R}$ defined as

$$\phi_p(x) = x(\log x - \log p) + (1-x)(\log(1-x) - \log(1-p)).$$

Note that ϕ_p is bounded by $2p/e$ and that the derivative of ϕ_p is given by

$$\phi_p'(x) = \log x - \log p - \log(1-x) + \log(1-p) = \log\left(1 + \frac{x-p}{p}\right) - \log\left(1 - \frac{x-p}{1-p}\right).$$

Recall that $\log(1+x) \sim x$ as $x \to 0$ and therefore, there exists a constant $C > 0$ and an integer $n_0 \in \mathbb{N}$ such that

$$|\phi_p'(x)| \leq C\sqrt{\frac{\log n}{n}}, \quad x \in [p - \sqrt{\log n/n}, p + \sqrt{\log n/n}], \ n \geq n_0. \tag{A.1}$$

Finally, note that, for any random variable X with $\mathbb{E}[X^2] < \infty$ and an independent copy $\widetilde{X}$,

$$\mathbb{E}[(X - \widetilde{X})^2] = \mathbb{E}[(X - \mathbb{E}[X]) - (\widetilde{X} - \mathbb{E}[\widetilde{X}]))^2]$$
$$= 2\mathrm{Var}(X) - 2\mathbb{E}[(X - \mathbb{E}[X])(\widetilde{X} - \mathbb{E}[\widetilde{X}])]$$
$$= 2\mathrm{Var}(X).$$

APPENDIX

Now let $\widetilde{B}_{n,p}$ be an independent copy of $B_{n,p}$. Then, in accordance with the previous observations,

$$\begin{aligned}
\operatorname{Var}(\phi_p(B_{n,p})) &= \frac{1}{2}\mathbb{E}[(\phi_p(B_{n,p}/n) - \phi_p(\widetilde{B}_{n,p}))^2] \\
&\leq \frac{1}{2}\mathbb{E}[(\phi_p(B_{n,p}/n) - \phi_p(\widetilde{B}_{n,p}))^2 \mathbb{1}_{\{|B_{n,p}-np|\leq\sqrt{n\log n}\}\cap\{|\widetilde{B}_{n,p}-np|\leq\sqrt{n\log n}\}}] \\
&\quad + (2p/e)^2 \mathbb{P}(|B_{n,p} - np| > \sqrt{n\log n}) \\
&\leq \frac{C^2 \log n}{2n}\mathbb{E}[(B_{n,p}/n - \widetilde{B}_{n,p})^2] + (2p/e)^2 \mathbb{P}(|B_{n,p} - np| > \sqrt{n\log n})
\end{aligned}$$

where the last bound holds by (A.1) and the mean value theorem.

Hence, $\mathbb{E}[(B_{n,p}/n - \widetilde{B}_{n,p})^2] = 2\operatorname{Var}(B_{n,p}/n) = 2p(1-p)/n$ and a standard Chernoff bound given in theorem A.1.1 yield

$$\operatorname{Var}(\phi_p(B_{n,p})) \leq \frac{C^2 p(1-p) \log n}{n^2} + 2(2p/e)^2 \exp\left(-2(\log n)^2\right) = \mathrm{O}\left(\frac{\log n}{n^2}\right).$$

This yields

$$\operatorname{Var}\left(B_{n,p}(\log(B_{n,p}/n) - \log p) + (n - B_{n,p})(\log(1 - B_{n,p}/n) - \log(1-p))\right) = \operatorname{Var}(n\phi_p(B_{n,p}/n))$$
$$= \mathrm{O}(\log n)$$

which is the assertion. □

Lemma A.2.2. *Let $p \in (0,1)$ be some constant and $B_{n,p}$ be binomial $B(n,p)$ distributed for $n \in \mathbb{N}$. Moreover, let $s \geq 1$ and $h : [0,1] \to \mathbb{R}$ be defined as $h(x) := x\log x$ with the convention $x\log x := 0$ for $x = 0$. Then, the following bounds hold as $n \to \infty$:*

$$\mathbb{E}\left[\log\left(\frac{B_{n,p}+1}{n+1}\right) - \log p\right] = \mathrm{O}\left(n^{-1/2}\right), \tag{A.2}$$

$$\mathbb{E}\left[B_{n,p}(\log(B_{n,p}+1) - \log(D_{n,p}))\mathbb{1}_{\{B_{n,p}\geq 1\}}\right] - 1 + \mathrm{O}\left(n^{-1}\right), \tag{A.3}$$

$$\|h(B_{n,p}/n) - \mathbb{E}[h(B_{n,p}/n)]\|_s = \mathrm{O}\left(n^{-1/2}\right), \tag{A.4}$$

$$\mathbb{E}[h(B_{n,p}/n) - h(p)] = \mathrm{O}(n^{-2/3}). \tag{A.5}$$

Proof. Proof of (A.2): The mean value theorem yields for all $\varepsilon \in (0,1)$

$$|\log(x) - \log(y)| \leq \varepsilon^{-1}|x - y|, \quad x, y \in [\varepsilon, 1].$$

Hence,

$$\left|\mathbb{E}\left[\log\left(\frac{B_{n,p}+1}{n+1}\right) - \log p\right]\right|$$
$$\leq \mathbb{E}\left[\left|\log\left(\frac{B_{n,p}+1}{n+1}\right) - \log p\right|\mathbb{1}_{\{B_{n,p}\geq np/2\}}\right] + \mathrm{O}\left(\log n\mathbb{P}(B_{n,p} < np/2)\right)$$
$$\leq \frac{2}{p}\mathbb{E}\left[\left|\frac{B_{n,p}+1-np-p}{n+1}\right|\right] + \mathrm{O}\left(\log n\mathbb{P}(B_{n,p} < np/2)\right).$$

The assertion follows because $\mathbb{E}[|(B_{n,p}-np)/\sqrt{np(1-p)}|]$ converges to the first absolute moment of the standard normal distribution (see [23, Theorem 4.2]) and a standard Chernoff bound given in theorem A.1.1 implies $\log n\mathbb{P}(B_{n,p} < np/2) = o(n^{-1/2})$.

Proof of (A.3): Note that $x \mapsto x(\log(x+1) - \log x)$ is bounded on $(0, \infty)$ and therefore,

$$\mathbb{E}\left[B_{n,p}(\log(B_{n,p}+1) - \log(B_{n,p}))\right]$$
$$= \mathbb{E}\left[B_{n,p}(\log(B_{n,p}+1) - \log(B_{n,p})\mathbb{1}_{\{B_{n,p} \geq np/2\}})\right] + \mathrm{O}\left(n\mathbb{P}(B_{n,p} < np/2)\right)$$
$$= 1 + \mathbb{E}\left[B_{n,p}(\log(1 + 1/B_{n,p}) - 1/B_{n,p})\mathbb{1}_{\{B_{n,p} \geq np/2\}})\right] + \mathrm{O}\left(n\mathbb{P}(B_{n,p} < np/2)\right)$$

which implies the assertion since $\log(1+x) - x = \mathrm{O}(x^2)$ as $x \to 0$ and the bound given in theorem A.1.1 yields $n\mathbb{P}(X_{n,p} < np/2) = o(n^{-1})$.

Proof of (A.4): First note that h is bounded on $[0,1]$ and that the derivative satisfies for all $\varepsilon > 0$

$$|h'(x)| \leq \log(1/\varepsilon) + 1, \quad x \in [\varepsilon, 1].$$

In particular, the mean value theorem implies

$$|h(x) - h(y)| \leq (\log(1/\varepsilon) + 1)|x - y|, \quad x, y \in [\varepsilon, 1]. \tag{A.6}$$

Let $\widetilde{B}_{n,p}$ be an independent copy of $B_{n,p}$. Then, Jensen's inequality and (A.6) yield

$$\|h(B_{n,p}/n) - \mathbb{E}[h(B_{n,p}/n)]\|_s^s$$
$$= \mathbb{E}[(\mathbb{E}[h(B_{n,p}/n) - h(\widetilde{B}_{n,p}/n)|B_{n,p}])^s]$$
$$\leq \mathbb{E}[(h(B_{n,p}/n) - h(\widetilde{B}_{n,p}/n))^s]$$
$$= \mathbb{E}[(h(B_{n,p}/n) - h(\widetilde{B}_{n,p}/n))^s \mathbb{1}_{\{B_{n,p}, \widetilde{B}_{n,p} \in [np/2, n]\}}] + \mathrm{O}(\mathbb{P}(B_{n,p} \leq np/2))$$
$$\leq (\log(2/p) + 1)^s \mathbb{E}[(B_{n,p}/n - \widetilde{B}_{n,p}/n)^s] + \mathrm{O}(\mathbb{P}(B_{n,p} \leq np/2))$$
$$\leq \left(\frac{\log(2/p) + 1}{\sqrt{n}}\right)^s (2\|(B_{n,p} - \mathbb{E}[B_{n,p}])/\sqrt{n}\|_s)^s + \mathrm{O}(\mathbb{P}(B_{n,p} \leq np/2)).$$

The assertion is derived from a standard Chernoff bound on $\mathbb{P}(X_{n,p} \leq np/2)$ that is given in theorem A.1.1 and $\|(B_{n,p} - \mathbb{E}[B_{n,p}])/\sqrt{n}\|_s \to \|Z\|_s$ where Z is $\mathcal{N}(0, p(1-p))$ distributed (cf. [23, Theorem 4.2] for details).

Proof of (A.5): It is sufficient to show that

(a) $h(p) - p\mathbb{E}[\log(B_{n,p}/n)\mathbb{1}_{\{B_{n,p} \geq 1\}}] = \mathrm{O}(n^{-2/3})$,

(b) $\mathbb{E}[h(B_{n,p}/n) - p\log(B_{n,p}/n)\mathbb{1}_{\{B_{n,p} \geq 1\}}] = \mathrm{O}(n^{-2/3})$.

For the first part note that

$$\left|h(p) - p\mathbb{E}[\log(B_{n,p}/n)\mathbb{1}_{\{B_{n,p} \geq 1\}}]\right|$$
$$= p\left|\mathbb{E}\left[\log\left(\frac{B_{n,p}}{np}\right)\mathbb{1}_{\{B_{n,p} \geq 1\}}\right]\right| + \mathrm{O}\left((1-p)^n\right)$$
$$= p\left|\mathbb{E}\left[\left(\log\left(1 + \frac{B_{n,p} - np}{np}\right) - \frac{B_{n,p} - np}{np}\right)\mathbb{1}_{\{B_{n,p} \geq 1\}}\right]\right| + \mathrm{O}\left((1-p)^n\right)$$
$$\leq p\left|\mathbb{E}\left[\left(\log\left(1 + \frac{B_{n,p} - np}{np}\right) - \frac{B_{n,p} - np}{np}\right)\mathbb{1}_{\{|B_{n,p} - np| \leq n^{2/3}\}}\right]\right|$$
$$+ (\log(np) + 1/p)\mathbb{P}(|B_{n,p} - np| > n^{2/3}) + \mathrm{O}\left((1-p)^n\right).$$

The upper bounds $\log(1+x) - x = \mathrm{O}(x^2)$ for $x \to 0$ and $\mathbb{P}(|X_{n,p} - np| > n^{2/3}) = o(n^{-1})$ by theorem A.1.1 imply

$$h(p) - p\mathbb{E}[\log(B_{n,p}/n)\mathbb{1}_{\{B_{n,p} \geq 1\}}] = \mathrm{O}(n^{-2/3}).$$

APPENDIX

In order to obtain bound (b), note that

$$\mathbb{E}[h(B_{n,p}/n) - p\log(B_{n,p}/n)\mathbb{1}_{\{B_{n,p}\geq 1\}}]$$
$$= \mathbb{E}\left[(h(B_{n,p}/n) - p\log(B_{n,p}/n))\mathbb{1}_{\{B_{n,p}\geq 1\}}\right] + O((1-p)^n)$$
$$= \frac{1}{\sqrt{n}}\mathbb{E}\left[\frac{B_{n,p}-np}{\sqrt{n}}\log\left(\frac{B_{n,p}}{n}\right)\mathbb{1}_{\{B_{n,p}\geq 1\}}\right] + O((1-p)^n)$$
$$= \frac{1}{\sqrt{n}}\mathbb{E}\left[\frac{B_{n,p}-np}{\sqrt{n}}\log\left(\frac{B_{n,p}}{n}\right)\mathbb{1}_{\{|B_{n,p}-np|\leq n^{2/3}\}}\right] + o\left(n^{-2/3}\right)$$
$$= \frac{1}{\sqrt{n}}\mathbb{E}\left[\frac{B_{n,p}-np}{\sqrt{n}}\log(p)\mathbb{1}_{\{|B_{n,p}-np|\leq n^{2/3}\}}\right]$$
$$+ \frac{1}{\sqrt{n}}\mathbb{E}\left[\frac{B_{n,p}-np}{\sqrt{n}}\log\left(1+\frac{B_{n,p}-np}{np}\right)\mathbb{1}_{\{|B_{n,p}-np|\leq n^{2/3}\}}\right] + o\left(n^{-2/3}\right).$$

Since $\log(1+x) = O(x)$ as $x \to 0$ and $\mathbb{E}[|(X_{n,p}-np)/\sqrt{n}|]$ converges to the first absolute moment of the $\mathcal{N}(0, p(1-p))$ distribution, the second summand is bounded by

$$\frac{1}{\sqrt{n}}\mathbb{E}\left[\frac{B_{n,p}-np}{\sqrt{n}}\log\left(1+\frac{B_{n,p}-np}{np}\right)\mathbb{1}_{\{|B_{n,p}-np|\leq n^{2/3}\}}\right] = O(n^{-5/6}).$$

For the first summand note that $\mathbb{E}[(B_{n,p}-np)/\sqrt{n}] = 0$ which implies

$$\frac{1}{\sqrt{n}}\mathbb{E}\left[\frac{B_{n,p}-np}{\sqrt{n}}\log(p)\mathbb{1}_{\{|B_{n,p}-np|\leq n^{2/3}\}}\right]$$
$$= -\frac{1}{\sqrt{n}}\mathbb{E}\left[\frac{B_{n,p}-np}{\sqrt{n}}\log(p)\mathbb{1}_{\{|B_{n,p}-np|> n^{2/3}\}}\right]$$
$$= O(\mathbb{P}(|B_{n,p}-np| > n^{2/3}))$$
$$= o(n^{-2/3}).$$

This yields the upper bound $\mathbb{E}[q(X_{n,p}/n) - p\log(X_{n,p}/n)\mathbb{1}_{\{X_{n,p}\geq 1\}}] = O(n^{-2/3})$ which combined with the first result implies the assertion. $\square$

The next lemma provides asymptotic results for the Poisson distribution that are needed for Poissonization in the analysis of the variance:

Lemma A.2.3. *For $\lambda > 0$ let N_λ follow the Poisson distribution $\Pi(\lambda)$. Then, the following bounds hold for all $\alpha, \beta > 0$ as $\lambda \to \infty$:*

$$\mathbb{E}[N_\lambda^\alpha] = O(\lambda^\alpha),$$
$$\mathbb{E}\left[N_\lambda^\alpha(\log N_\lambda)^\beta\right] = O\left(\lambda^\alpha(\log \lambda)^\beta\right).$$

Proof. The upper bound on $\mathbb{E}[N_\lambda^\alpha]$ is derived in three steps:

(1) For $\alpha \in \mathbb{N}$ the assertion is shown by an induction on α and the fact that, for every $n \in \mathbb{N}_0$,

$$\mathbb{E}\left[\prod_{i=0}^n (N_\lambda - i)\right] = \sum_{k=n+1}^\infty e^{-\lambda}\frac{\lambda^k}{k!}\prod_{i=0}^n (k-i)$$
$$= \lambda^{n+1}\sum_{k=n+1}^\infty e^{-\lambda}\frac{\lambda^{k-(n+1)}}{(k-(n+1))!}$$
$$= \lambda^{n+1}.$$

The induction relies on the identity

$$\mathbb{E}[N_\lambda^\alpha] + \sum_{k=0}^{\alpha-1}\binom{\alpha}{k}\mathbb{E}[N_\lambda^k](-\alpha)^{\alpha-k} = \mathbb{E}[(N_\lambda-\alpha)^\alpha] \le \mathbb{E}\left[\prod_{i=0}^{\alpha-1}(N_\lambda-i)\right] = \lambda^\alpha.$$

(2) For $\alpha \in (0,1)$ note that $x \mapsto x^\alpha$ is concave on $[0,\infty)$ and therefore, by Jensen's inequality

$$\mathbb{E}[N_\lambda^\alpha] \le (\mathbb{E}[N_\lambda])^\alpha = \lambda^\alpha.$$

(3) For $\alpha \in (1,\infty) \cap \mathbb{N}^c$ note that $x \mapsto x^{\alpha/\lceil\alpha\rceil}$ is concave on $[0,\infty)$ which yields

$$\mathbb{E}[N_\lambda^\alpha] \le (\mathbb{E}[N_\lambda^{\lceil\alpha\rceil}])^{\alpha/\lceil\alpha\rceil}$$

and the assertion follows from part (1).

The upper bound for $\mathbb{E}\left[N_\lambda^\alpha(\log N_\lambda)^\beta\right]$ is derived from the following decomposition:

$$\begin{aligned}\mathbb{E}\left[N_\lambda^\alpha(\log N_\lambda)^\beta\right] &= \mathbb{E}\left[N_\lambda^\alpha(\log N_\lambda)^\beta \mathbb{1}_{\{N_\lambda \le \lambda^{\alpha+1}\}}\right] + \mathbb{E}\left[N_\lambda^\alpha(\log N_\lambda)^\beta \mathbb{1}_{\{N_\lambda > \lambda^{\alpha+1}\}}\right] \\ &\le (\alpha+1)^\beta (\log \lambda)^\beta \mathbb{E}[N_\lambda^\alpha] + \mathbb{E}\left[N_\lambda^\alpha(\log N_\lambda)^\beta \mathbb{1}_{\{N_\lambda > \lambda^{\alpha+1}\}}\right] \\ &= \mathrm{O}(\lambda^\alpha(\log\lambda)^\beta) + \mathbb{E}\left[N_\lambda^\alpha(\log N_\lambda)^\beta \mathbb{1}_{\{N_\lambda > \lambda^{\alpha+1}\}}\right],\end{aligned}$$

where the last step holds since $\mathbb{E}[N_\lambda^\alpha] = \mathrm{O}(\lambda^\alpha)$. Hence, it is sufficient to show that

$$\mathbb{E}\left[N_\lambda^\alpha(\log N_\lambda)^\beta \mathbb{1}_{\{N_\lambda > \lambda^{\alpha+1}\}}\right] = \mathrm{O}(\lambda^\alpha).$$

The upper bound $n^\alpha(\log n)^\beta \le C_{\alpha\beta} n^{3\alpha/2}$ for a sufficiently large constant $C_{\alpha\beta}$ and all $n \in \mathbb{N}_0$ yields

$$\begin{aligned}\mathbb{E}\left[N_\lambda^\alpha(\log N_\lambda)^\beta \mathbb{1}_{\{N_\lambda > \lambda^{\alpha+1}\}}\right] &\le C_{\alpha\beta}\mathbb{E}\left[N_\lambda^{3\alpha/2}\mathbb{1}_{\{N_\lambda > \lambda^{\alpha+1}\}}\right] \\ &\le C_{\alpha\beta}\sqrt{\mathbb{E}[N_\lambda^{3\alpha}]\mathbb{P}(N_\lambda > \lambda^{\alpha+1})}\end{aligned}$$

where the last inequality holds by the Cauchy-Schwarz inequality. Together with the previous result $\mathbb{E}[N_\lambda^{3\alpha}] = \mathrm{O}(\lambda^{3\alpha})$ and Markov's inequality one obtains

$$\mathbb{E}\left[N_\lambda^\alpha(\log N_\lambda)^\beta \mathbb{1}_{\{N_\lambda > \lambda^{\alpha+1}\}}\right] = \mathrm{O}(\lambda^\alpha)$$

and the assertion follows. □

Bibliography

[1] Cesaratto, E. and Vallée, B. (2015) Gaussian Distribution of Trie Depth for Strongly Tame Sources. *Combinatorics, Probability and Computing* **24**, 54–103.

[2] Clément, J., Flajolet, Ph. and Vallée, B. (2001) Dynamical sources in information theory: a general analysis of trie structures. Average-case analysis of algorithms (Princeton, NJ, 1998). *Algorithmica* **29**, 307–369.

[3] Clément, J. Fill, J. A., Flajolet, Ph. and Vallée, B. (2009) The number of symbol comparisons in QuickSort and QuickSelect. *Lecture Notes in Comput. Sci.* **5555**, 750–763.

[4] Clément, J., Nguyen Thi, T. H. and Vallée, B. (2013) A general framework for the realistic analysis of sorting and searching algorithms. Application to some popular algorithms. *30th International Symposium on Theoretical Aspects of Computer Science* **20**, 598–609.

[5] de la Briandais, R. (1959) File searching using variable length keys, in *Proceedings of the AFIPS Spring Joint Computer Conference.* AFIPS Press, Reston, Va., 295-298.

[6] Devroye, L. (1982) A note on the average depth of tries. *Computing* **28**, 367–371.

[7] Devroye, L. (1984) A probabilistic analysis of the height of tries and of the complexity of triesort. *Acta Informatica* **21**, 229–237.

[8] Devroye, L. (1992) A study of trie-like structures under the density model. *Ann. Appl. Probab.* **2**, 402–434.

[9] Devroye, L. (2002) Laws of large numbers and tail inequalities for random tries and PATRICIA trees. *J. Comput. Appl. Math.* **142**, 27–37.

[10] Devroye, L. (2005) Universal asymptotics for random tries and PATRICIA trees. *Algorithmica* **42**, 11–29.

[11] Devroye, L., Szpankowski, W. and Rais, B. (1992) A note on the height of suffix trees *SIAM J. Comput.* **21**, 48–53.

[12] Drmota, M., Janson, S. and Neininger, R. (2008) A functional limit theorem for the profile of search trees. *Ann. Appl. Probab.* **18**, 288–333.

[13] Drmota, M. and Szpankowski, W. (2011) The expected profile of digital search trees. *J. Combin. Theory Ser. A* **7**, 1939–1965.

[14] Fill, J.A. and Kapur, N. (2004) The Space Requirement of m-ary Search Trees: Distributional Asymptotics for $m \geq 27$. Invited paper, Proceedings of the 7th Iranian Statistical Conference, 2004. Available via http://www.ams.jhu.edu/~fill/papers/periodic.pdf

[15] Flajolet, Ph., Jacquet, P.,Mahmoud, H.M. and Régnier, M. (2000) Analytic variations on bucket selection and sorting. *Acta Inform.* **36**, 735–760.

[16] Flajolet, Ph., Roux, M. and Vallée, B. (2010) Digital trees and memoryless sources: from arithmetics to analysis. *21st International Meeting on Probabilistic, Combinatorial, and Asymptotic Methods in the Analysis of Algorithms (AofA'10)*, Discrete Math. Theor. Comput. Sci. Proc., AM, Assoc. Discrete Math. Theor. Comput. Sci., Nancy, 233–260.

[17] Flajolet, Ph. and Sedgewick, R. (1986) Digital search trees revisited. *SIAM J. Comput.* **15** 748–767.

[18] Flajolet, Ph. and Sedgewick, R. (2009) *Analytic combinatorics*. Cambridge University Press, Cambridge.

[19] Fuchs, M., Hwang, H.K. and Zacharovas, V. (2014) An analytic approach to the asymptotic variance of trie statistics and related structures. *Theoret. Comput. Sci.* **527**, 1–36.

[20] Fuchs, M., Lee, C.-K. and Prodinger, H. (2012) Approximate counting via the poisson-laplace-mellin method *23rd Intern. Meeting on Probabilistic, Combinatorial, and Asymptotic Methods for the Analysis of Algorithms (AofA'12)*, 13–28.

[21] Givens, C. R. and Shortt, R. M. (1984) A class of Wasserstein metrics for probability distributions. *Michigan Math. J.* **31**, 231–240.

[22] Gusfield, D. (1997) *Algorithms on Strings, Trees, and Sequences*, Cambridge University Press, Cambridge.

[23] Gut, A. (2009) *Stopped random walks*, second edition, Springer, New York.

[24] Hubalek, F. (2000) On the variance of the internal path length of generalized digital trees-the mellin convolution approach. *Theoret. Comput. Sci.* **242**, 143–168.

[25] Hun, K. and Vallée, B. (2014) Typical depth of a digital search tree built on a general source. *ANALCO14—Meeting on Analytic Algorithmics and Combinatorics*, 1–15.

[26] Hwang, H.-K., Nicodème, P.,Park, G. and Szpankowski, W. (2008) Profile of tries. *LATIN 2008: Theoretical informatics*, 1–11.

[27] Jacquet, Ph. and Régnier, M. (1986) Trie partitioning process: limiting distributions. *CAAP '86 (Nice, 1986)*, 196–210.

[28] Jacquet, Ph. and Régnier, M. (1988) Normal limiting distribution of the size and the external path length of tries. Technical Report RR-0827, INRIA-Rocquencourt.

[29] Jacquet, Ph. and Régnier, M. (1988) Normal limiting distribution of the size of tries. *Performance '87 (Brussels, 1987)*, 209–223, North-Holland, Amsterdam.

[30] Jacquet, Ph. and Régnier, M. (1989) New results on the size of tries. *IEEE Trans. Inform. Theory* **35**,203–205.

[31] Jacquet, Ph. and Szpankowski, W. (1989) Analysis of Digital Tries with Markovian Dependency. *Computer Science Technical Reports*. Report 89-906, Purdue University. Available via http://docs.lib.purdue.edu/cstech/772

[32] Jacquet, Ph. and Szpankowski, W. (1991) Analysis of digital tries with Markovian dependency. *IEEE Trans. Information Theory*, **37**, 1470–1475.

[33] Jacquet, Ph. and Szpankowski, W. (1994) Autocorrelation on words and its applications: analysis of suffix trees by string-ruler approach. *J. Combin. Theory Ser. A* **66**, 237–269.

[34] Jacquet, Ph. and Szpankowski, W. (1995) Asymptotic behavior of the Lempel-Ziv parsing scheme and [in] digital search trees. Special volume on mathematical analysis of algorithms. *Theoret. Comput. Sci.* **144**, 161–197.

[35] Jacquet, Ph. and Szpankowski, W. (1998) Analytical Depoissonization and Its Applications, *Theoretical Computer Science*, 201, 1–62.

[36] Jacquet, Ph. and Szpankowski, W. (2011) Limiting distribution of lempel ziv'78 redundancy *Information Theory Proceedings (ISIT), 2011 IEEE International Symposium on*, 1509–1513.

[37] Jacquet, P., Szpankowski, W. and Tang, J. (2001) Average profile of the Lempel-Ziv parsing scheme for a Markovian source. Mathematical analysis of algorithms. *Algorithmica* **31**, 318–360.

[38] Janson, S. (2012) Renewal theory in the analysis of tries and strings *Theoret. Comput. Sci.* **416**, 33–54.

[39] Janson, S. and Neininger, R. (2008) The size of random fragmentation trees. *Probab. Theory Related Fields* **142**, 399–442.

[40] Kirschenhofer, P. and Prodinger, H. (1988) Further results on digital search trees. Thirteenth International Colloquium on Automata, Languages and Programming (Rennes, 1986). *Theoret. Comput. Sci.* **58**, 143–154.

[41] Kirschenhofer, P., Prodinger, H. and Szpankowski, W. (1989) On the variance of the external path length in a symmetric digital trie. Combinatorics and complexity (Chicago, IL, 1987). *Discrete Appl. Math.* **25**, 129–143.

[42] Kirschenhofer, P., Prodinger, H. and Szpankowski, W. (1994) Digital search trees again revisited: the internal path length perspective. *SIAM J. Comput.* **23**, 598–616.

[43] Kirschenhofer, P., Prodinger, H. and Szpankowski, W. (1996) Analysis of a Splitting Process Arising in Probabilistic Counting and Other Related Algorithms, *Random Structures & Algorithms*, 9, 379–401.

[44] Knape, M. and R. Neininger (2014) Pólya urns via the contraction method. *Comb. Probab. & Comput.*, **23**, 1148–1186.

[45] Knessel, C. and Szpankowski, W. (2009) On the average profile of symmetric digital search trees *Online J. Anal. Comb.* **4**, 14.

[46] Knuth, D.E. (1998) *The Art of Computer Programming, Volume III: Sorting and Searching*, Second edition, Addison Wesley, Reading, MA.

[47] Leckey, K., Neininger, R. and Sulzbach, H. (2014) Analysis of radix selection on Markov sources. *Proceedings of the 25th International Conference on Probabilistic, Combinatorial and Asymptotic Methods for the Analysis of Algorithms (AofA'14)*, 253–264

[48] Leckey, K., Neininger, R. and Szpankowski, W. (2013) Towards More Realistic Probabilistic Models for Data Structures: The External Path Length in Tries under the Markov Model. *Proceedings ACM-SIAM Symp. Disc. Algo. (SODA)*, 877–886.

[49] Louchard, G.(1987) Exact and asymptotic distributions in digital and binary search trees. *RAIRO Inform. Théor. Appl.* **21**, 479–495.

[50] Louchard, G. and Szpankowski, W. (1995) Average profile and limiting distribution for a phrase size in the lempel-ziv parsing algorithm *Information Theory, IEEE Transactions on* **41**, 478–488.

[51] Mahmoud, H.M. (1992) *Evolution of Random Search Trees*, John Wiley & Sons, New York.

[52] Major, P. (1978) On the invariance principle for sums of independent identically distributed random variables. *J. Multivariate Anal.* **8**, 487–517.

[53] Neininger, R. (2001) On a multivariate contraction method for random recursive structures with applications to Quicksort. Analysis of algorithms (Krynica Morska, 2000). *Random Structures Algorithms* **19**, 498–524.

[54] Neininger, R. and Rüschendorf, L. (2004) A general limit theorem for recursive algorithms and combinatorial structures. *Ann. Appl. Probab.* **14**, 378–418.

[55] Neininger, R. and Rüschendorf, L. (2004) On the contraction method with degenerate limit equation. *Ann. Probab.* **32**, 2838–2856.

[56] Neininger, R. and Sulzbach, H. (2012) On a functional contraction method. *Ann. Probab.*, to appear.

[57] Rachev, S.T. and Rüschendorf, L. (1995) Probability metrics and recursive algorithms. *Adv. in Appl. Probab.* **27**, 770–799.

[58] Rais, B., Jacquet, Ph. and Szpankowski, W. (1993) Limiting distribution for the depth in PATRICIA tries. *SIAM J. Discrete Math.* **6**, 197–213.

[59] Rösler, U. (1991) A limit theorem for "Quicksort". *RAIRO Inform. Théor. Appl.* **25**, 85–100.

[60] Rösler, U. (1992) A fixed point theorem for distributions. *Stochastic Process. Appl.* **42**, 195–214.

[61] Rösler, U. (1999) On the analysis of stochastic divide and conquer algorithms. Average-case analysis of algorithms (Princeton, NJ, 1998). *Algorithmica* **29**, 238–261.

[62] Rösler, U. and Rüschendorf, L. (2001) The contraction method for recursive algorithms. *Algorithmica* **29**, 3–33.

[63] Schachinger, W. (1995) On the variance of a class of inductive valuations of data structures for digital search. *Theoret. Comput. Sci.* **144**, 251–275. Special volume on mathematical analysis of algorithms.

[64] Szpankowski, W. (1988) Some results on V-ary asymmetric tries *J. Algorithms* **9**, 224–244.

[65] Szpankowski, W. (1991) A characterization of digital search trees from the successful search viewpoint. *Theoret. Comput. Sci.* **85**, 117–134.

[66] Szpankowski, W. (1993) A generalized suffix tree and its (un)expected asymptotic behaviors. *SIAM J. Comput.* **22**, 1176–1198.

[67] Szpankowski, W. (2001) *Average Case Analysis of Algorithms on Sequences*, John Wiley, New York.

[68] Vallée, B. (2001) Dynamical sources in information theory: fundamental intervals and word prefixes, *Algorithmica* **1-2**, 262–306.

[69] Ziv, J. and Lempel, A. (1978) Compression of individual sequences via variable-rate coding *IEEE Trans. Inform. Theory* **24**, 530–536.

[70] Zolotarev, V. M. (1976) Approximation of the distributions of sums of independent random variables with values in infinite-dimensional spaces. (Russian.) *Teor. Veroyatnost. i Primenen.* **21**, 741–758. Erratum *ibid* **22** (1977), 901. English transl. *Theory Probab. Appl.* **21**, 721–737; *ibid.* **22**, 881.

[71] Zolotarev, V. M. (1977) Ideal metrics in the problem of approximating the distributions of sums of independent random variables. (Russian.) *Teor. Veroyatnost. i Primenen.* **22**, 449–465. English transl. *Theory Probab. Appl.* **22**, 433–449.

I want morebooks!

Buy your books fast and straightforward online - at one of the world's fastest growing online book stores! Environmentally sound due to Print-on-Demand technologies.

Buy your books online at
www.get-morebooks.com

Kaufen Sie Ihre Bücher schnell und unkompliziert online – auf einer der am schnellsten wachsenden Buchhandelsplattformen weltweit! Dank Print-On-Demand umwelt- und ressourcenschonend produziert.

Bücher schneller online kaufen
www.morebooks.de

OmniScriptum Marketing DEU GmbH
Heinrich-Böcking-Str. 6-8
D - 66121 Saarbrücken
Telefax: +49 681 93 81 567-9

info@omniscriptum.com
www.omniscriptum.com

Printed by Books on Demand GmbH, Norderstedt / Germany